THE INVENTIONS OF DAEDALUS

A Compendium of Plausible Schemes

DAVID E. H. JONES

W. H. FREEMAN AND COMPANY

Oxford and New York

W. H. Freeman and Company Limited
20 Beaumont Street, Oxford, England OX1 2NQ
41 Madison Avenue, New York, New York 10010

Library of Congress Cataloging in Publication Data
Jones, David E. H.
The inventions of Daedalus.
A compilation and embellishment of 129 selected
Daedalus columns from the magazine, *New Scientist*.
Includes index.
1. Inventions–Miscellanea. I. *New Scientist*. II. Title.
T20.J66 500 81–19605
ISBN 0–7167–1412–4 AACR2
ISBN 0–7167–1413–2 (pbk.)

Set by Interprint Ltd., Malta
Printed in the United States of America.

3 4 5 6 7 8 9 V B 0 8 9 8 7 6 5 4

Contents

Introduction

About Daedalus

This book is a compilation and embellishment of 129 selected 'Daedalus' schemes from the magazine *New Scientist*. Daedalus, whose column appears every week on Ariadne's page of the magazine, is some sort of re-incarnation of his ancient Greek prototype. His remit is to devise inventions rather in the spirit of the wax-and-feather wings of his illustrious predecessor. That is to say, they must be ingenious and novel, but should raise questions in the mind of the reader. They should ideally fall in that uneasy no man's land between the clearly possible and the clearly fantastic.

I have been writing the 'Daedalus' column for about 17 years, and it has long been an ambition of mine to assemble a selection of them into a book. With well over 800 columns to choose from, there was no lack of material. But its organization posed problems. A mere scissors-and-paste assembly would have been quite indigestible. Besides, many of the schemes as originally printed were seriously cramped by the space-limitations of the *New Scientist* slot, and I liked the idea of being able to expand them. So the entries in this book do not always follow the original text exactly. I have felt free to expand some sections for clarity, to include notions I was forced to cut out of the originals, to merge together schemes which originally had to be split for presentation on two successive weeks, and to perform various other acts of editing. In particular, the vast majority of past Daedaluses began with a traditional formula 'My (adjective) friend Daedalus . . .' where the adjective was a fanciful word describing the topic to be discussed. The endless repetition of this phrase soon became unbearably tedious. So I cut it out of the entries in this book, and stopped using it in the column.

Daedalus's flights of inventive fantasy are always launched from the solid ground of scientific reality. Indeed, despite my best efforts, some 17% of them are later seriously suggested in one way or another, patented, made to work, or even turn out to have been done already! I've even made some of them work myself, for scientific television programmes. So I was highly delighted to learn that at one stage the Physics Department of the University of Manchester began giving its students selected Daedalus columns as design studies. The students enjoyed the challenge but the results were difficult to mark. Accordingly, for those who wish to probe deeper into the scientific background, I have included entries 'From Daedalus's notebook' presenting some of the calculations behind his schemes. I have reworked all of them for this book, sometimes uncovering small errors, and on one occasion at least an enormous blunder. So the numerical conclusions do not always agree exactly with those originally printed. Other random remarks, including some intriguing instances of Daedaluses-come-true, are marshalled under 'Daedalus comments'.

For many years my vague intentions of producing a Daedalus book seemed blocked by the problem of diagrams and cartoons. Where could I find an artist with the technical insight needed to do justice to Daedalus's fantasies? The publishers finally solved the problem by inveigling me into doing all the drawings myself. The bizarre consequences are scattered throughout the book.

About DREADCO

DREADCO, whose activities crop up in most of the entries in this book, is the Daedalus Research, Evaluation And Development Corporation. It is the contract-research organization of which Daedalus is Research Director. It was founded in 1967, a couple of years after the Daedalus column itself, as a brilliant suggestion of Edward Wheeler, then Technology Editor of *New Scientist*. With the resources of DREADCO behind him, Daedalus is able to develop schemes of truly industrial complexity, and the firm has gone from strength to strength ever since. Its philosophy was outlined by Daedalus in the first DREADCO Annual Report, published in *New Scientist*, 24 December 1970:

The corporation's profitability continues to rest upon its research contracts. Practically all this income is ploughed into research and development activity with the customary disregard of what our clients think they want. As you know, our success depends on the fact that nearly all major scientific advances have been made while looking for something else, or following up curious observations. To this end we continue to encourage the utmost scientific irresponsibility in our staff: varying research goals at whim, trying crazy random experiments, and so on. Not only does this policy give our laboratories an exciting atmosphere lacking from their more expensively endowed but narrowly commercial industrial counterparts, but it aids the recruitment of the most creative workers, and regularly leads to quite unpredicted developments of great value to us or our customers . . .

As you should know by now, I have no patience with the fiction that a company exists for the benefit of its shareholders, who in any case tend to be a transient, anonymous, quick-profit-minded and scientifically illiterate crew. DREADCO succeeds the way it does only because it is run solely for the satisfaction and entertainment of the staff, particularly me.

DREADCO is not primarily a manufacturing organization. It markets new inventions on some fairly modest pilot scale, mainly for evaluation. For all that, it seems over the years to have accumulated a considerable degree of goodwill in the industrial community at large. *New Scientist* has received a number of serious enquiries about licencing arrangements, etc., for DREADCO's inventions. DREADCO has even been threatened with legal action for patent infringement! As long as such things continue to happen from time to time, I shall know that the firm is being run on the right lines.

DAVID JONES

*Newcastle upon Tyne,
May 1981*

Dyed in the woolly

There appears to be a strange gap in the colourful biochemistry of the animal kingdom: no animal has green fur. There must be considerable camouflage advantage in green coloration for pasture animals, but Nature seems unable to achieve this advance. Daedalus has therefore been seeking ways of modifying animal fur, and recalls in this connection the famous scientific verification of the arsenic-poisoning of Napoleon by detection of the metal in samples of his hair. Biologists have already suggested that the repeated skin-shedding of creatures like snakes is a secondary excretion mechanism. So Daedalus reckons that animals and humans also use their detachable surface features — nails, hair and fur — as a similar dump for dangerous metals. Indeed, the process presumably improves with practice, as small doses of arsenic can gradually build up a considerable tolerance. Daedalus therefore proposes to encourage metal-rejection in sheep by gradually dosing them with compounds of unhealthy but colourful metals: copper (blue), nickel (green), cobalt (pink) and so on. Suitable biochemical investigations should reveal the compounds most effectively rejected into the wool. Not only might this lead to better camouflaged green sheep (or more usefully, green cows and horses) capable of surviving in regions now barred to them by predators, but genuine permanently dyed-in-the-wool fabrics would at last be available. Even better, by alternating the dose metal during the growth of the fibres, multicoloured tabby sheep could be produced. They would be ideal for the manufacture of Fair Isle pullovers, Harris Tweed, and similarly subtly-patterned garments. Daedalus is now devising metal-containing synthetic polymers, whose incorporation into animal feed could well lead directly to dyed polymer/wool staple in which the rival fibres would be chemically merged for instant weaving into those popular hybrid fibres.

(*New Scientist*, 18 January 1968)

From Daedalus's notebook

The uptake of arsenic into hair seems a good model system to start work on. The war-gas Lewisite is a typical arsenical. It reacts with the thiol groups in certain enzymes, in this manner:

$$\begin{array}{c} \boxed{}\begin{array}{l}-SH\\-SH\end{array} + \begin{array}{l}Cl\\Cl\end{array}>As-CH=CH-Cl \\ \text{Enzyme} \qquad\qquad \text{Lewisite} \\ \downarrow \\ \boxed{}\begin{array}{l}-S\\-S\end{array}>As-CH=CH-Cl \quad + 2\,HCl \\ \text{Deactivated enzyme} \end{array}$$

It's the thiol groups in keratin that are responsible for the uptake of arsenic into hair and nails. That double-bonded organic side-chain on the Lewisite molecule is very reminiscent of the multiple double-bonded structures of most organic dyes. So it would seem entirely feasible to synthesize compounds of this nature:

With luck they would if ingested bind to hair in humans or to the fur in animals. And since most organic dyes are so strongly coloured that only micrograms are needed to produce quite strong colour, there's a good chance that a dose too small to poison a sheep might still colour its wool quite dramatically. Furthermore, since we can vary the dye moiety at will, we could make the creatures any colour — green, purple, even Dayglo orange (very useful for finding them in snowdrifts).

What pattern would result? It's very unlikely that sheep wool grows uniformly all over the animal all the time. It probably grows slowly in some areas and faster in others, interspersed with periods of rest (like human hair). So the slow-growing areas should accumulate the dye in much greater density than the faster-growing ones, and a fascinating latent pattern should be revealed. If the active regions are closely spaced, some sort of tabby should result; or there may merely be multicoloured blotches. But once the pattern of growth has been established, phased feeding with different colour-formers could be timed to coincide with the growth of selected areas. That way really gorgeous multicoloured animals could be produced. I wonder if they would be self-conscious?

Fizzy solids

A new children's sweet contains compressed carbon dioxide occluded in sugar granules, and crackles alarmingly in the mouth as the sugar dissolves to release the gas. Daedalus sees vast implications in the principle. He emphasizes that very small bubbles exert great pressures on the gas inside them anyway, so by dispersing fine bubbles in liquids and letting them set under external pressure, really energetic foamed solids could be made. Thus DREADCO's pressure-foamed soap, when rubbed wet on the skin, delivers a vigorous vibro-massage as it dissolves to release thousands of tiny gas eruptions. The characteristic rude noise accompanying this discharge can be 'tuned' by varying the bubble size, making possible a foamed detergent whose built-in ultrasonic agitation does away with washing-machines, and an ultrasonic toothpaste giving both high polish and a really intense tingling sensation. Daedalus also began occluding compressed oxygen in Kendal Mint Cake to provide an energy-rich food for divers, as the released oxygen should be absorbed by the gut and save them from having to breathe. But a perfectly balanced sample, capable of being metabolized completely in its own oxygen, turned out rather alarmingly to be high explosive. Edible dynamite could fill a valuable dual role in the supplies for explorers of rough country, but less appetizing oxy-foamed explosives are being developed.

Another DREADCO team is working on pressure-foamed fillings for chairs and sofas. Recalling the deadliness of the fumes evolved by burning polyurethane foam, Daedalus will pressure his new sofa filling with the combustion-inhibitors used in some fire-extinguishers. As the sofa softens in a fire, its millions of tiny bubbles will burst and release a great blast of gas to blow out the flames, accompanied by a huge raspberry of sound to raise the alarm. Indeed, houses liberally furnished in the new foam would probably never catch fire in the first place, as a dropped match or cigarette would be puffed out as soon as it melted its first hole.

(*New Scientist*, 14 September 1978)

From Daedalus's notebook

The typical solubilities of gases in liquids are in the range 0.005–0.1% by weight at 1 atmosphere pressure. Since the solubility goes up directly with the applied pressure, 500 atmospheres should dissolve gases at 2.5–50% by weight. Further gas can be forced into a liquid by introducing fine bubbles. The pressure inside a bubble exceeds that outside by $2\gamma/r$, where γ is the surface tension of the liquid and r is the radius of the bubble. Typical liquid surface tensions are of the order of $\gamma = 0.05\,\text{N m}^{-1}$, so that a bubble $10^{-8}\,\text{m}$ across ($r = 5 \times 10^{-9}\,\text{m}$) would compress the contained gas to $p = 2 \times 0.05/(5 \times 10^{-9}) = 2 \times 10^7$ $\text{N m}^{-2} = 200$ atmospheres.

Probably the best way to make such tiny bubbles is to saturate the liquid with bigger bubbles and then raise the pressure to compress them. But a combination of pressure-solubility and bubble-occlusion seems well able to produce a mixture containing an appreciable, or even dominant, proportion of the gas. Then you let the liquid set. Molten plastics and melted sugar are obvious candidates, especially if suitable additives can be found having a high inherent solubility for the chosen gas; and pressures of a few hundred atmospheres appear adequate. Note that sugar is combustible: $C_{12}H_{22}O_{11}$ (342 g) + $12O_2$ (192 g) = $12CO_2 + 11H_2O$; so sugar containing only 36% of dissolved oxygen could be perfectly metabolized or burnt in its own gas-content.

Daedalus comments

The children's sweet referred to in this column is General Foods' 'Space Dust', which had recently been marketed in Britain at the time. It is made (*Chemical Technology*, July 1978, p. 446) by dissolving carbon dioxide gas under 40 atmospheres pressure in a warm sugar-syrup, and letting it set. The resulting solid contains about $2\,\text{cm}^3\,\text{g}^{-1}$ of carbon dioxide occluded in the crystal lattice, and about $4\,\text{cm}^3\,\text{g}^{-1}$ as tiny bubbles: it's the bubbles which give it the fizz.

The product has an interesting history (*The Economist*, 26 May 1979, p. 114), dating from the early 1960s. The original — and very ingenious — idea was to create a solid which could be dissolved in water to make a fizzy carbonated drink. This didn't work — either they couldn't get enough carbon dioxide into the sugar, or else it just bubbled out without dissolving when the crystals were put in water. But the stuff did have this strange fizzing effect when dissolved in the mouth, so in 1962 General Foods patented it, and then tried to sell licences for the process to sweet-makers. But they asked too high a price to attract any takers, and not till 1978, when the patents were nearing expiry, did General Foods take the plunge itself and market 'Space Dust'.

The idea of making a high explosive by mixing oxygen intimately with a combustible material is entirely practical too. The Simplon Tunnel was blasted with an explosive made by dipping sticks of charcoal in liquid oxygen. But here the high concentration of the oxidizer was obtained by low temperature rather than high pressure.

Failure to maintain firm hold of DREADCO's pressure-foamed soap

Facoder

The current interest in the vision-telephone prompts Daedalus to wonder what human purpose will be served by such an instrument. No verbal advantage is gained by seeing your conversant's face; but it does let you pick up subtle indications that, e.g., he's about to finish his remarks, or that he's angry, or that he's lying. This last factor is presumably the main interest of businessmen. But to convey such elementary points by means of a continuous TV picture, requiring for its transmission as much bandwidth as a thousand audio channels, is technical overacting on the grand scale. In this connection Daedalus recalls the 'Vocoder', a gadget which coded speech into a sort of phonetic telegraph code that could be transmitted very economically. At the far end, the code was used to drive a synthetic voice which was quite understandable but lacked the character and timbre of the original speaker. Impressed by the ease with which a good caricaturist can represent a face and its mood in a few lines only, he is devising the facial equivalent of the Vocoder. The DREADCO 'Facoder' extracts from a TV-camera output a cartoon of the speaking face, by advanced computer pattern-recognition methods. This is then represented by a simple 'identikit' code and transmitted down the telephone cable to be displayed at the receiver. Such a simple signal, changing only 20 times a second or so, would take up very little bandwidth and could easily squeeze in alongside the voice transmission. This elegant gadget will enable the whole telephone network to be equipped with vision, without the ghastly expense of total rewiring with video cable!

To optimize the necessary computer software, DREADCO psychologists are studying non-verbal communication to decide just how the face codes its signals. They are studying the works of Walt Disney, filmed interviews with politicians covering a broad spectrum of dishonesty and evasiveness, and mimed exchanges between subjects separated by frosted glass of varying degrees of translucency. From the results, the Facoder-codes will be written, enabling the machine to transmit pure facial information without irrelevant details of lighting, background, etc. It may be possible to 'recognize' the identity of your conversant from his caricature, but most of its visual content will be emotional. You will seem to converse with a flexible cartoon-face which will nod agreement, or express doubt, hostility, warmth or whatever in almost pure form — rather like Alice's Cheshire cat. An intensity control will enable you to amplify slight irritation into towering rage, say, or fractional hesitancy into total bewilderment: a useful feature in talking to stiff-upper-lip public-school types. Conversely, an attenuator for reducing continental passions to decent English proportions will greatly help the European traffic. Thus many of the disconcerting barriers of national character and style will be overcome. The world's various communities will be put into much truer and more harmonious communication than is possible by direct face-to-face dialogue.

Even more beneficial will be the use of the Facoder to break the infuriating communication-barrier between road-users. A Facoder-camera pointing at the driver's face and driving display screens on the front and rear of his vehicle, would revolutionize traffic safety if universally adopted. All the frustrations of incommunicado motoring would melt away. At a glance you could 'read' the vehicles around you as motivated by mania, distraught with worry, or drifting absentmindedly — and then act accordingly. The blunderer could apologize via his crestfallen caricature, visibly defusing the anger of the wronged; and the mute regret on the face of the mis-firing banger would atone to those stuck behind it. The furious emotions which seem to be stimulated by normal driving should be replaced by instinctive polite facial interchange.

But the true significance of Daedalus's invention probably lies in its ability to bridge the biggest communication gap of them all — that between man and machine. In controlling a large chemical plant, say, you need to react fast and accurately to the things that matter. A vast bank of dials is a ridiculously crude way of discovering what's going on, and a pile of computer print-out is even worse. By contrast, we all learnt at an early age to read the many nuances of human expression, and developed powerful instinctive skills in learning the right tricks to keep Big Daddy happy. So DREADCO engineers are devising a Facoder system whereby the control panel of a chemical plant will not be a bank of meters but a screen displaying one or more Facoder faces. Variations from the norm will each impose a characteristic expression on the faces: happiness and buoyancy if output rises; various subtle types of unease if certain units are below par; rank fear if catastrophe threatens. The control engineers will instantly read the state of the plant, and their normal human learning instincts will guide them into optimizing it as never before. Such display methods might even develop emotional sensitivity in the more spiritually stunted engineers, and would do much to hasten 'technology with a human face'. On the other hand, it might also greatly increase the number of workers falling in love with their job.

(*New Scientist*, 25 October and 1 November 1973)

GLOBAL OPTIMISATION

MANAGER   FEED PUMP No1   FEED PUMP No2

CONDENSER   COLUMN No1   COLUMN No2   REBOILER

HEAT EXCH No2   CRACKER

9

Some domestic delights

In Daedalus's view, the plight of housewives has been only cursorily relieved by technology. While more and more chromium-plated devices are available to perform domestic chores, their fundamental nature has not changed. Daedalus now presents some results of a radical reconsideration of housework. The whole paraphernalia of cooking, eating and washing-up can be drastically simplified, and for a start the inefficient and dirty oven could be replaced by a simple rod-like electric element to be pushed into the cake, chicken or whatever, to heat it from the *inside*. A disposable insulating wrapper would prevent heat and vapour losses. Even better, why not manufacture foods in individual portions? Daedalus suggests as a universal food-module a small filled cylinder carrying its own electric heater, a sort of plug-in sausage in fact. All separate items of crockery would be replaced by his conveyor table, a slowly moving belt toroidally moulded to divide into shallow pans serving as saucepans, plates, etc. Sloppy foods like soup or porridge would be heated in these at one end, and eaten as they passed the central section; the belt would be continuously washed up by a revolving scrubber as it returned under the table.

Most of the problems of dusting and cleaning arise from entropy effects spreading the stuff out, and these can easily be reversed by providing some potential-well for it to fall into. Thus if a tray of glycerine or treacle were placed anywhere in a room while the dust was kept on the move by a powerful air-blower, sooner or later every speck would land on, and be trapped in, the sticky liquid. Again a slightly sloping floor which could be periodically vibrated would cause dislodged rubbish to accumulate in the lowest corner for easy removal. Alternatively, a moving carpet like the conveyor-table with a continuous carpet-cleaner underneath would work, but special arrangements with castors and string would be needed to hold the furniture in place.

(*New Scientist*, 5 January 1967)

Daedalus comments

I am happy to record that, following my advocacy, the electric sausage was indeed invented. *New Scientist* (9 September 1971, p. 577) reported this excerpt from British Patent 1 228 914 granted to Anstalt Euroresearch: 'An article of solid food, such as a sausage, is then put in the cooker with one end in each of the electrolyte pools and a current passed through the electrodes, electrolyte and sausage circuit.' Thus the resistance of the sausage is used to heat it, a development I had also pioneered (*New Scientist*, 17 April 1969). But don't prick it with a metal fork while it's cooking!

Some while ago Daedalus began experimenting with a novel dusting technique which only needed a tray of treacle and a powerful electric fan. The stirred air circulates all the dust; ultimately every particle must by chance land on the treacle, where it sticks and sinks in. After many dustings the treacle becomes solid with dust, whereupon you put it through the mangle, regenerating the treacle and producing a handy felt mat into the bargain. This process is an example of the 'Maxwell Demon' principle of separating objects by letting them go one way and not the other. It won't work for molecules (as Maxwell originally suggested) but is fine for larger objects like dust-particles, and Daedalus is devising several such demonic devices. Thus his humane fly demon is attached to a window or a ventilator. Whenever a fly, in its crazy wanderings, approaches the demon, sensors open the demon's door and let it out. Flies outside are refused entry, however, so the machine inevitably clears the house of those insects with a random component in their flight (i.e. all of them). The same principle will work even better with surface creatures like mice and cockroaches. Sooner or later curiosity or blind wanderings bring them to the demon's door, when they are politely allowed out.

A more selective demon will employ sophisticated optical, acoustic and electronic sensors feeding a self-optimizing 'learning' program in its decision-making microcomputer. It could then be 'taught' to admit a selected species by repeated association of the opening command with presentation of appropriate specimens. Set up in the wall of an enclosure it would steadily and humanely accumulate the desired species inside, while ejecting those not required. So Daedalus now proposes demonic fishing. This simply needs a pipeline extending from shore and terminating in a demon programmed to admit cod, herring, or whatever. A steady flow of the desired species should be filtered effortlessly from the eddying shoals and passed down the pipe for shore processing.

(*New Scientist*, 1 October 1970)

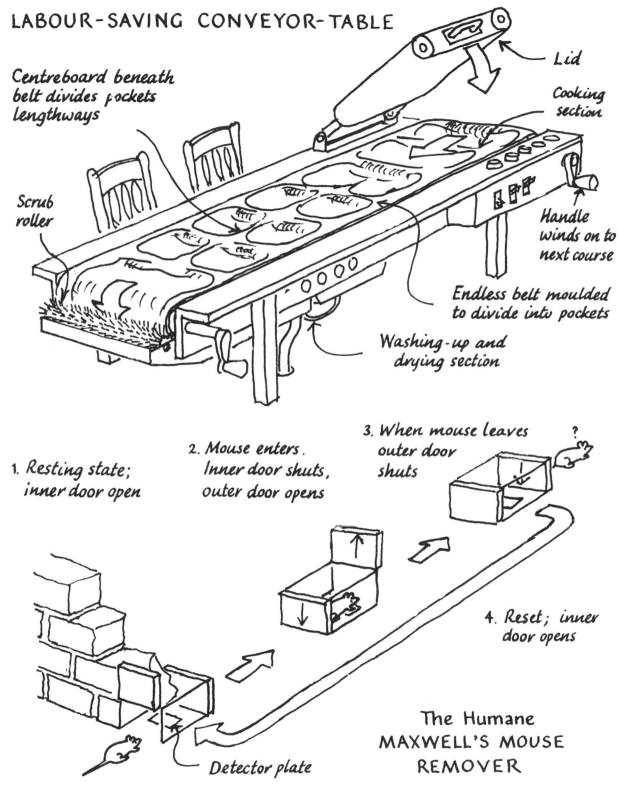

From Daedalus's notebook

LABOUR-SAVING CONVEYOR-TABLE

Centreboard beneath belt divides pockets lengthways

Lid

Cooking section

Scrub roller

Handle winds on to next course

Endless belt moulded to divide into pockets

Washing-up and drying section

1. Resting state; inner door open

2. Mouse enters. Inner door shuts, outer door opens

3. When mouse leaves outer door shuts

4. Reset; inner door opens

The Humane MAXWELL'S MOUSE REMOVER

Detector plate

Digging for electricity

The 'homopolar generator' theory of the Earth's magnetic field asserts that as the molten iron in the Earth's core moves by convection through that field, electric currents are generated on the dynamo principle, and these in turn sustain the field. Daedalus sees in these vast Earth-currents the answer to our energy problem, if only he can sink electrodes deep enough to tap them. Ordinary drilling gives out at a few kilometres depth. Daedalus, however, recalls that all rocks are really plastic and the Earth is in approximate hydrostatic equilibrium throughout. That's why oil deep down is under pressure, and oil companies pump heavy mud down their boreholes to counterbalance that pressure. Suppose, says Daedalus, that a 10-km borehole were filled not with mud, but with the far denser mercury. The hydrostatic pressure at the bottom would be about 13 000 atmospheres, far in excess of the local rock-pressure. So the rock would gradually yield — indeed it might yield quite rapidly, as the temperature at that depth may exceed $400\,^\circ$C. So the mercury would force its way downwards: and with steady topping-up from the surface the process would accelerate indefinitely.

Now all solids must conduct at a high enough temperature (by thermal excitation of their electrons). So Daedalus hopes that his mercury electrode-drill will encounter dynamo-currents in the superheated rocks only a few tens of kilometres down, and not have to penetrate the 1000 km or so into the Earth's iron core itself. Furthermore, as the liquid drill encounters these ever-higher temperatures, he will be able to top it up with cheaper, higher-melting alloys from Wood's metal to, finally, molten iron. He will tap into the gigantic Earth-dynamo at several points of differing polarity so as to exploit the maximum voltage-difference available. This may only be 100 volts or so, but the internal resistance of the whole Earth must be so low that billions of amps could be drawn out without noticeably shorting it out. The new earthly power-source will effortlessly solve the energy-crisis with no pollution or environmental damage. But Daedalus wonders if the Friends of the Earth will really approve.

(*New Scientist*, 14 July 1977)

Last week Daedalus revealed his scheme to sink boreholes down to where the currents run which sustain the Earth's magnetic field: and then to extract this free electricity through molten-metal electrodes. He now points out that enough current could easily be drawn from the vast Earth-dynamo to heat the metal-electrode columns fiercely. When they had become so hot that the surrounding melted rock became conducting too, they would no longer be needed. A self-sustaining white-hot current-bearing column, like a huge Nernst filament, would connect the surface with the depths: a sort of tame electric volcano, or 'electrano'. It would deliver copious heat to the surface, derived both from electrical heating and convection of molten rock from the fiery depths.

Convectively mined deep rock would be of great geological interest and (since Daedalus suspects that the heavier elements like gold, platinum, palladium, etc., must mainly have sunk to these depths over geological time) of huge financial potential. The downgoing rock currents could usefully carry away all our rubbish, including the real nasties like radioactive and carcinogenic waste. And electranoes will be scientifically useful too. The seismic creaks and groans of the whole Earth must be piezoelectrically transmitted into its conducting layers by the strained hot rock. The resulting a.c. will be amplified by the same homopolar mechanism which sustains the Earth currents and geomagnetic field; so 'listening' to the a.c. ripple on the output of an electrano would give powerful geophysical insights. It might enable earthquakes to be predicted, or even pre-emptively set off by back injection of appropriate a.c. resonances which, amplified by the homopolar mechanism, would induce piezoelectric resonant failure in the overloaded rock. Similarly, telegraph signals fed into one electrano would be amplified for reception by others all over the world. Indeed, they might even be 'broadcast' on the Earth's magnetic field to be read on magnetic compasses everywhere.

(*New Scientist*, 21 July 1977)

ELECTRANO TERMINAL

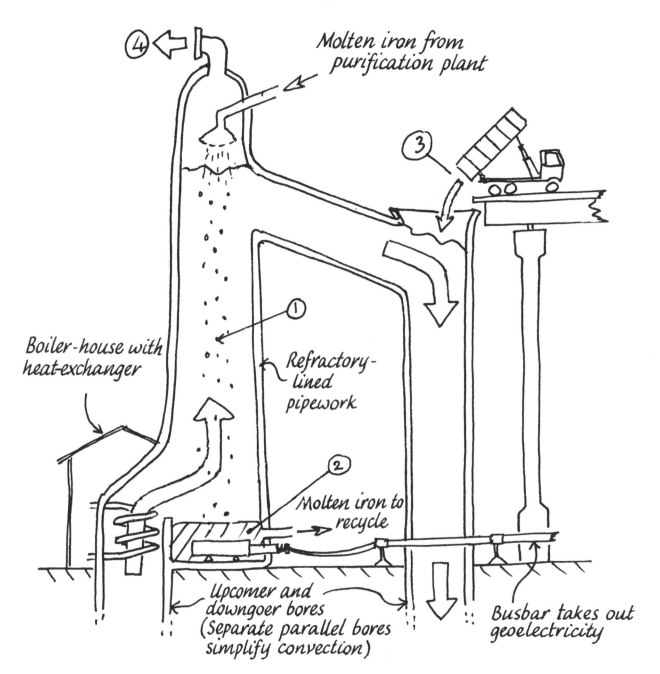

Molten iron from purification plant

Boiler-house with heat-exchanger

Refractory-lined pipework

Molten iron to recycle

Upcomer and downgoer bores (Separate parallel bores simplify convection)

Busbar takes out geoelectricity

Numbered details:
1. Molten metal (eg. iron) sprayed down ascending lava in counter-current fashion to extract heavy metals
2. Molten-iron electrode takes out current and also cathodically strips ionic metals from lava
3. Rubbish is fed into downgoing lava
4. Gas take-off. Gases entrained in deep-earth lava may include useful methane, etc.

Cleaning by electric traction

In chemical engineering terms, washing-up is absurdly wasteful: it takes a vast amount of water to shift only a little dirt. Laundering and human washing are even more prodigal, and many industrial processes are worse still. And since each particle of dirt is held in suspension by a firmly adhering coating of detergent molecules, that expensive ingredient goes down the drain too. Seeking some economy measure, Daedalus recalled electrodeposition, the painting technique whereby a suspension of paint droplets in water is caused to deposit on metal car bodies by the passage of an electric current. Similarly, he reasons, the dirt in his washing-up water could be removed by electrodepositing it, paint-fashion, on a suitable electrode. As the particles coalesced into a film, their surface area would drop, expelling the surface-adsorbed detergent molecules. Clean, sudsy, detergent-laden water would be released for the next washing-up. So Daedalus is inventing everlasting washing-up bowls, permanent baths and laundry-tubs, etc., to exploit this principle. The dirt introduced into the water is simply plated out again by electrodeposition, and one charge of detergent and clean water lasts for ever!

Furthermore, electrodeposition at one electrode might well be accompanied by the converse process, electro-suspension, at the other. Perhaps a dirty fork or grubby shirt could be stripped of its grime by silent electrical attraction rather than human or mechanical elbow-grease. Daedalus is working on a hygienic electric bath, in which the user as one electrode is anodically stripped of his surface grime, which winds up plated onto a counter-electrode near the plug-hole. The electrical extraction of surface dirt might give absolutely the cleanest wash ever experienced. It would certainly give that tingling-fresh sensation, particularly if high voltages or ultrasonic vibration proved necessary. But what to do with the dirt-plated electrodes? The rich, greasy, fibre-laden deposit might adhere to them as firmly as wet paint. So one idea is to use the stuff as a paint, and bake it on. But the appealing idea of a combined public wash-house and car-body paint shop, with the vehicles as electrodes taking up the grime of collective ablution, would probably fail from the limited colour-range of the products. Daedalus's electrodeposited dirt will probably be scraped off and sold, prosaically, for growing mushrooms.

(*New Scientist*, 20 July 1978)

From Daedalus's notebook

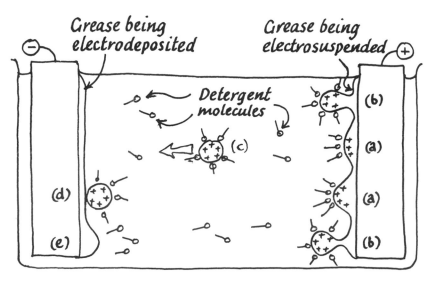

The positive electrode carries the grease to be removed. Its surface tension, already lowered by adsorption of detergent molecules (a), is opposed by the surface-repulsion of the applied positive charge; ultimately positively-charged grease droplets (b) are repelled from the electrode and travel in suspension (c) towards the negative electrode. On touching it (d) the positive charge is neutralized. The droplet merges into the existing grease-film (e) and the reduction in surface area expels the adsorbed detergent molecules which diffuse back to the positive electrode

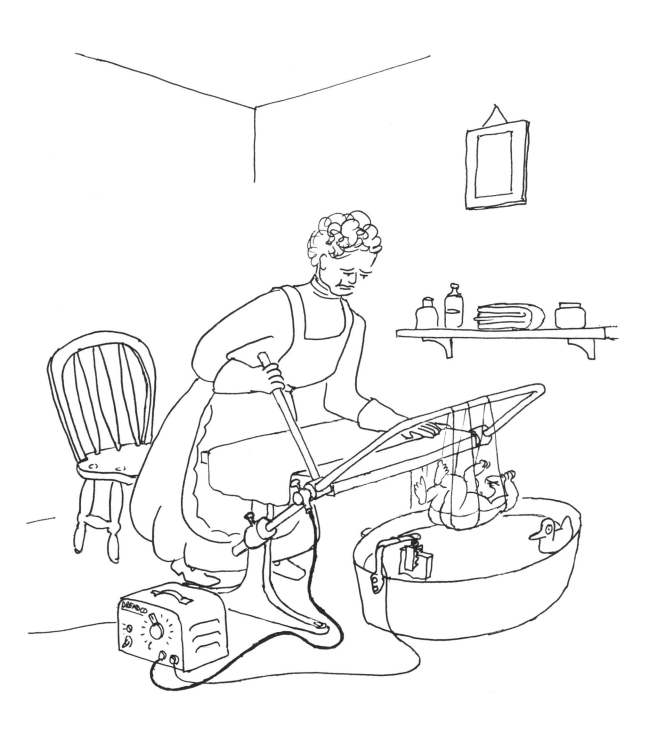

*The DREADCO electrosuspensory baby-bath cleans baby immaculately all over, without
exposing tender young skin to rough handling or abrasive scrubbing*

Heat pumps and hot pants

Daedalus is pondering the dilemma of keeping warm. Current fashion seems to demand the minimum amount of thin, tight-fitting clothing — quite the worst recipe for heat-retention. So he is devising a new and thermodynamically elegant solution — heat-pump clothes. If your thin jeans, for example, took in ambient heat at (say) 10 °C, and delivered it to your skin at 35 °C, they would feel warmer than the most extravagently furry trousers. Yet for this modest temperature gradient, a heat-pump could easily transfer inwards as heat at least 10 times the energy needed to power it. Daedalus's scheme draws heavily on modern hollow-fibre heat-exchanger technology. DREADCO polymer technologists are extruding hollow fibres from vibrating nozzles to introduce regularly spaced kinks and constrictions into them. In particular they are looking for asymmetric kinks to act as one-way valves. Once a hollow-fibre can be drawn with alternate fine constrictions and one-way valves in it, it will be extruded in an atmosphere of fluorocarbon vapour. The fluorocarbon, chosen to have the proper volatility for a heat-pump medium, will partially condense within the hollow fibre.

The fluid-filled fibre will be woven into the looped pile of a towelling-like cloth. With careful weaving the alternate valves and constrictions will occur in the plane of the cloth, with the loops of tubing extending out of the plane on either side. Imagine, says Daedalus, a pair of trousers made from this cloth. Every time the wearer moves, the fibre will flex and each loop will change in volume, rather like a Bourdon pressure-gauge. Accordingly each loop will act as a tiny peristaltic pump. The fluorocarbon vapour held in the outer loops will be compressed and pushed through the central one-way valves into the inner loops, where it will condense to liquid and deliver its latent heat to the wearer's skin. When the wearer reverses his movement, the outer loops will increase in volume again, allowing the liquid in the inner loops to leak back into them through the fine central constrictions. There it will evaporate and take in more latent heat from the outside air. Thus every movement of the wearer drives the distributed heat-pump of his trousers. For every calorie thus expended as movement, he will receive up to 10 calories pumped inwards through his trousers! DREADCO's brilliant garments will effortlessly solve the comfort/fashion dilemma. Quite ordinary activity, like walking about or even breathing, will keep us warm as toast and save kilowatts of expensive central heating. Indeed, since heat-pump clothes could pump back into the body more energy than it actually expends, there could be curious metabolic consequences. Joggers in the new garments, for example, might find themselves putting on weight!

(*New Scientist*, 28 February 1980)

From Daedalus's notebook

Thermodynamic aspects of heat-pump clothing. The gain of a heat-pump, (heat evolved)/(work expended), is given by $\alpha = T_o/(T_o - T_i)$ where T_o is the output temperature and T_i the input temperature. Putting $T_o = 35\,°C = 308\,K$, and $T_i = 10\,°C = 283\,K$, we find $\alpha = 12.3\,J/J$. In practice this theoretical output can't be achieved, but even 10 joules of heat for every joule of work is a good rate of exchange. The best refrigerant is probably Freon-114 (dichlorotetrafluoroethane), boiling-point 4 °C. But for arctic trousers to be worn in extreme conditions, chlorotrifluoroethylene (b.p. $-$ 28 °C) would be better.

Forming the tubing. As with most fibre-extrusion work experience is the only sure guide. But with luck a nozzle that gives a sharp kick backwards will produce a fine constriction:

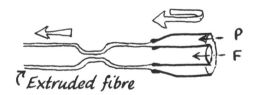

Extruded fibre

while one that gives a sharp forward kick will give a one-way valve:

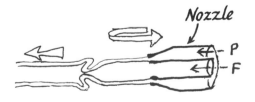

P = polymer
F = fluorocarbon

With piezoelectric actuators all sorts of strange wobbles can be imposed on the extruders, so something ought to work!

Action of the garments. What pumping-efficiency can we expect from loop-woven towelling? Not much, I'm afraid. Only quite sharp flexing of the substrate (e.g. at creases, joints like knee and hip, etc.) will alter the curvature and volume of the outer loops very much. A better approach is to weave the loops so that they have to bend sharply at the top; a kink will develop there. Very little flexing of the cloth will then suffice to move this kink along the tube, or open or close it, making a true peristaltic pump. And angular movements transferred from the inner loops, due to the cloth moving against the skin, will be even more efficient. Since such movements will occur over wide areas as the wearer moves about, this will pump much more heat than Bourdon-gauge flexing, and more uniformly too. It won't matter that the kinks will open, close, and move around quite irregularly; the one-way valves will ensure that pumping occurs consistently in the right direction.

A further advantage. It's a common reflex to rub any area you feel is chilly. With heat-pump garments this would flatten the outer loops momentarily and produce massive pumping rates. So even a static wearer in a pretty cold environment could still keep very warm by this instinctive action. Still, heat-pump gloves and socks, which will get a lot of flexing and cover extremities very liable to chill, are probably better garments than trousers for the pilot study.

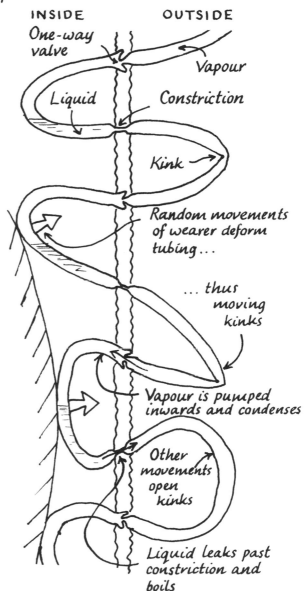

Cross-section of heat-pump garment, showing hollow-fibre 'pile'

INSIDE   OUTSIDE

One-way valve

Vapour

Liquid

Constriction

Kink →

Random movements of wearer deform tubing...

... thus moving kinks

Vapour is pumped inwards and condenses

Other movements open kinks

Liquid leaks past constriction and boils

Deadening the wake

Daedalus has been musing on the deplorable fact that travel is 0% efficient. Energy is continually put into driving the car, ship, plane or whatever along, and yet it finishes with no more energy than at its start. All the consumed energy has been degraded to heat along its path. Contemplating this disturbing conclusion, Daedalus recognized that the major source of loss is fluid resistance, creating the vast swirls of eddying air or water left behind by the vehicle. It occurred to him that if, on these waves and eddies, there could be superimposed a set of exactly opposite ones, complete calm would result. Unfortunately it would be far too difficult to calculate the pattern of turbulence which must be set up ahead of a vehicle in order that its passage should create just that equal and opposite disturbance to cancel it. The obvious answer is to fit a set of rear sensors to detect residual turbulence behind the vehicle, and use them to signal ahead to the eddy-generators in front. An appropriate self-optimizing control-system would soon 'learn' how to keep the aft-sensors reading practically zero. It would be a fairly simple problem in feedback control with time-delay round the loop (for the craft takes time to catch up with and cancel the eddies created ahead of it).

At first Daedalus was puzzled by the implications of this scheme. His anti-turbulence ship, or whatever, uses up more energy than before, for it must power its eddy-generators. Yet no energy is dissipated into the fluid medium, which is left calm and undisturbed after the ship's passage. Where does all the energy go? Daedalus's conclusion is that the eddy-machine in fact accelerates the ship: that pattern of vortices and waves which exactly cancels the turbulence of the vessel also and necessarily pulls it along. Furthermore, the absence of a wake means that this mode of propulsion requires almost no energy. All the power put into creating turbulence ahead of the vehicle is reabsorbed in its passage! Thus enormous economies are possible. Daedalus's vehicles, propelled by their servo-controlled flaps and fans up ahead, would glide effortlessly along leaving peace behind them. The aircraft's slipstream, the ship's wake, the car's disturbance would all be generated ahead of them and smoothly absorbed into power as they passed. Even the noise could be cancelled and put to good use.

(*New Scientist*, 15 August 1968)

Daedalus comments

> Big whirls have little whirls
> That feed on their velocity;
> And little whirls have lesser whirls,
> And so on to viscosity.

This epigram by L. F. Richardson, which sums up his classic paper of 1920 on atmospheric eddies, reminds us not to be taken in by the doleful assumption that mechanical losses in a process are always irretrievably degraded to heat. The scrambling of ordered mechanical movement right down to ultimate thermal molecular disorder often occurs in many stages, the parcels of coherent order getting smaller and smaller at each successive stage. But the Second Law of Thermodynamics permits us to rescue useful mechanical energy from every stage except the last, truly thermal one. In this scheme I propose recovering useful energy from the first few stages of fluid mechanical loss: big waves and macroscopic eddies. Imagine a film, taken from the air, of a ship moving through the water; and now imagine that film shown *backwards*. The bow-waves and wake-turbulence would be seen to converge on the vessel and be cancelled by its motion as they hit it. It would in effect be travelling through a particular form of choppy sea and using the wave-energy to propel itself. Now provided we don't consider processes on the molecular level, all macroscopic mechanical behaviour is reversible. So that imaginary backwards film depicts a genuinely possible event. The eddy-generators ahead of my proposed anti-turbulence ship would have to be optimized to create just the converging pattern of waves and eddies shown in the reversed film. Only viscosity and skin-friction, the true degraders of fluid energy to heat, cannot be reclaimed in this way, so these would still exact their toll. In obedience to the Second Law of Thermodynamics, the anti-turbulence ship would leave a calm but slightly warmer sea in its imperceptible wake.

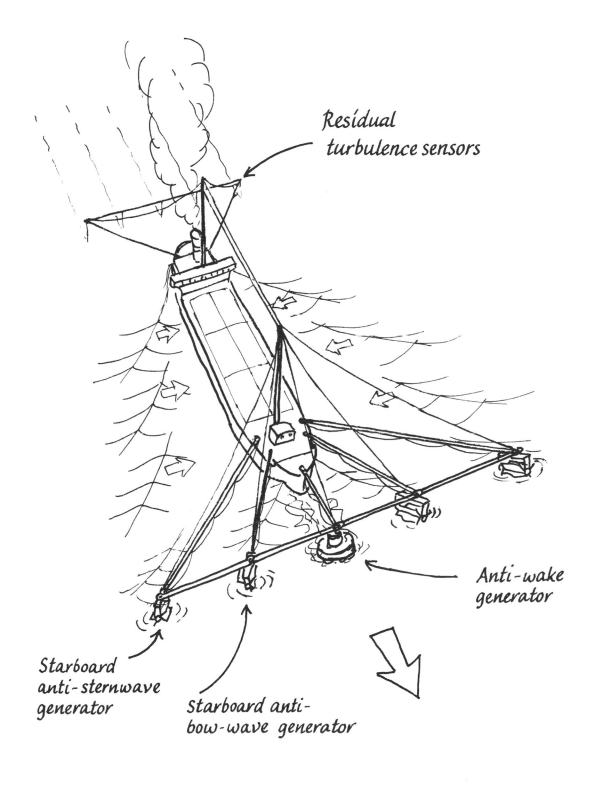

Residual
turbulence sensors

Anti-wake
generator

Starboard
anti-sternwave
generator

Starboard anti-
bow-wave generator

Dithergas

Daedalus has a chemical explanation for the mass-incompetence of elderly organizations so ably documented by Parkinson. The unconscious emission of 'pheromones' affecting the mood of others (e.g. sex-attractants, and the fear-substances which can spread panic or rouse the watchdog against the apprehensive postman) suggests to him that the almost tangible atmosphere of hopelessness in, e.g. old labour exchanges and welfare halls, is a chemical effluvium from the staff and clientele. Even young, dynamic recruits are quickly demoralized in such environments, by what must be a chemical defence evolved by those low in social status to deflate the aggression of higher-ranking individuals. Daedalus intends to isolate this elixir of incompetence and hopelessness from the air of Civil Service departments, chambers of commerce of fading seaside resorts, Liberal Party forward-planning committees and the like. He surmises that like other pheromones it may be a very simple chemical, but has not been identified before because of the sudden dithering ineptitude that afflicts chemists who accidentally synthesize it. But alert to the peculiar hazards of the enterprise, DREADCO should soon be marketing the new 'Dithergas' (Regd) in concentrated form as the ideal humane riot-control agent.

Being one of the body's own products, it cannot have dangerous side-effects, and its smell is slight if infinitely depressing. Under the impact of Dithergas the stoutest apostles of social justice will falter, beset by imbecilic doubts of their capacity to fulfil a task suddenly become bafflingly complex and obscure. In the grip of defeatism they will begin to evolve the resolve-sapping pheromone themselves, and soon the wildest demonstration will disperse dejectedly like a vicarage fête in a thunderstorm. DREADCO will also offer small aerosols of Dithergas as the *perfect* foil for airline hijackers — provided the pilot himself keeps his personal oxygen-mask on!

But the possible applications of Dithergas go far beyond its simple use as a riot-control agent. As one of the pheromones which stabilize the many subtleties of the social order, it could be used to sabotage or manipulate that order in many cunning ways. Thus it would be welcomed by executives eager to sabotage the departments of their rivals, thwarted Romeos seeking to weaken the resolve of haughty beauties, committee-men anxious to damp down opposition to their pet schemes, and military commanders bent on undermining enemy drive and efficiency. Indeed, Dithergas seems the absolutely perfect weapon for an undeclared war. Countries secretly blanketed with the nearly odourless gas would succumb inexplicably to vague misgivings and defeatist bumbling. But the idea alarms Daedalus — it explains so plausibly the present state of Britain and the resurgence of Germany and Japan!

Fortunately, there must be an antidote. Strong evolutionary pressures must have stimulated the production of blood-hormones neutralizing the effects of Dithergas — indeed, muses Daedalus, it is presumably just this unconscious 'chemical warfare' which controls the emergence of leaders and followers in all human societies. So he is seeking the antidote to Dithergas in the bloodstreams of those confident, exasperating eager-beavers who are immune to the effluvia of gormlessness generated in unconscious self-defence by their cringing underlings. DREADCO is offering free medical checkups to dynamic businessmen of all industries, obtaining thereby blood-samples and measures of individual resistance to Dithergas (information which, incidentally, should enable DREADCO's Stockmarket Investment Department to place its funds with uncommon insight). Once identified, the Dithergas-antidote should be ideal for revitalization and countering depression, not to mention national regeneration. But Daedalus is getting worried about selling such key biochemicals freely in a myopically competitive society. Once the secret is out, everyone may try to dither his rivals while boosting himself, and the whole social order (which these pheromones have evolved to stabilize) may be upset. It may be better to keep the knowledge safely within DREADCO, using Dithergas to clobber rival organizations which seem close to learning the secret.

(*New Scientist*, 26 March and 2 April 1970)

'. . . the sudden dithering ineptitude that afflicts chemists who accidentally synthesize it'

Slippery shipping

There is a special anti-fouling paint for ships that never quite dries, so that barnacles and marine organisms find it too shifty for adhesion. DREADCO's Paint Laboratories once compounded too fluid a version of such a paint. It slowly flowed down the ship's sides and dripped off the keel. Inspired by this happy accident, DREADCO chemists are now devising other such fluid-flow paints, to automate the expensive chore of maintenance painting. Applied to a house, the new paint would be continuously pumped to a horizontal spray-pipe along the roof-ridge. It would pour over the roof, forming a thick, putty-like layer flowing slowly down the tiles. Guttering under the eaves would transfer it to the walls, where its soggy downward progress would continue (special deflectors would stop it obscuring the windows). At ground level, the viscous sheet would enter receiving gutters for recycling via a clean up unit back to the roof.

This thick, self-healing, endlessly-renewed layer of paint would eliminate one of the chief annoyances of domestic ownership. Like pitch or silly putty (which also flow in this way) it would not be tacky to the touch. Its high viscosity may ensure even flow without rippling, but filtering and recycling it will give grave pumping problems. A melting or solvent-thinning step may be required. Even so, continuous automatic repainting will be a boon not only for houses but also for bridges, industrial plant and even ships. Learning from previous experience, Daedalus will give his new fluid-flow anti-fouling paint exactly the same density as seawater, so it will neither fall to the bottom of the sea nor float to the top. It will be pumped over the bows of the ship, and will slowly make its way by viscous drag around to the stern, where it will be pumped on board again for recycling. It may revolutionize marine engineering. For surface tension pulls a liquid surface molecularly smooth; furthermore such a surface, if soft and yielding, will damp out small eddies before they can grow. So ships so painted may imitate the dolphin, and cleave the water not by turbulent shear but in true laminar flow — cutting the power needed to maybe 10% of its previous value!

(*New Scientist*, 8 January 1981)

From Daedalus's notebook

What viscosity do we want for our ever-flowing paint? A film of liquid, viscosity η, density ρ, and thickness x, runs down a vertical wall with a mean velocity $\bar{v} = \rho g x^2/3\eta$, i.e. $\eta = \rho g x^2/3\bar{v}$. We want the flow down say 10 m of house elevation to take a month to a year — say 10^7 or 10^8 seconds, i.e. $\bar{v} = 10^{-6} - 10^{-7}\,\mathrm{m\,s^{-1}}$. With x about 1 mm, this puts η in the range $10^4 - 10^5\,\mathrm{N\,s\,m^{-2}}$: typical of resins and softish waxes.

With such tiny velocities, flow cannot possibly be turbulent. Even so, we may need proper control of surface tension and thixotropy to prevent slow laminar instabilities (tear-drops, rippling, etc.).

What economy can we expect if fluid-flow paint enables ships to operate in laminar rather than turbulent flow? According to R. G. Morgan (*Science News*, Vol. 40, 1956, p. 96) skin resistance for turbulent flow is:

$$R_{TF} = 0.455\,A(\log Re)^{-2.58}$$

whereas for laminar flow it is:

$$R_{LF} = 1.339\,A(Re)^{-0.5}$$

where Re is Reynold's no., $vl\rho/\eta$, and A is a normalized drag force, $0.5\,\rho v^2$ Newtons per square metre of wetted surface. Taking a modest vessel of length $l = 20\,\mathrm{m}$, $v = 5\,\mathrm{m\,s^{-1}}$ in water of $\rho = 1000\,\mathrm{kg\,m^{-3}}$ and $\eta = 10^{-3}\,\mathrm{N\,s\,m^{-2}}$, $R_{TF} = 27\,\mathrm{N\,m^{-2}}$ and $R_{LF} = 1.7\,\mathrm{N\,m^{-2}}$. Even allowing for wave-making losses pushing both of these upwards, there is easily a factor of 10 to be had!

Incidentally, we have this viscous goo flowing from bow to stern and offering hundreds of square metres of surface to rapidly flowing seawater, and then being recovered and purified for recycling. This is about the most ideal set-up imaginable for contacting thousands of tons of seawater for chemical extraction. Suppose we incorporate suitable complexing reagents in the paint. Then on its slow journey from bow to stern it could be made to pick up magnesium and bromine, maybe cobalt and mercury (both getting scarce), perhaps even gold. They could easily be stripped from the paint during the filtering and recycling process. Such elements are extremely dilute in seawater, but processing so many thousands of tons of it so effortlessly might still bring in a useful profit.

MARINE VISCOUS-PAINT SYSTEM

Block diagram

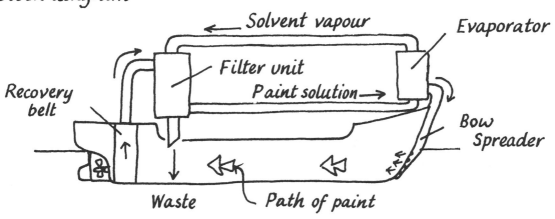

Solvent vapour

Evaporator

Filter unit

Paint solution →

Recovery belt

Bow Spreader

Waste

Path of paint

Detail of stern

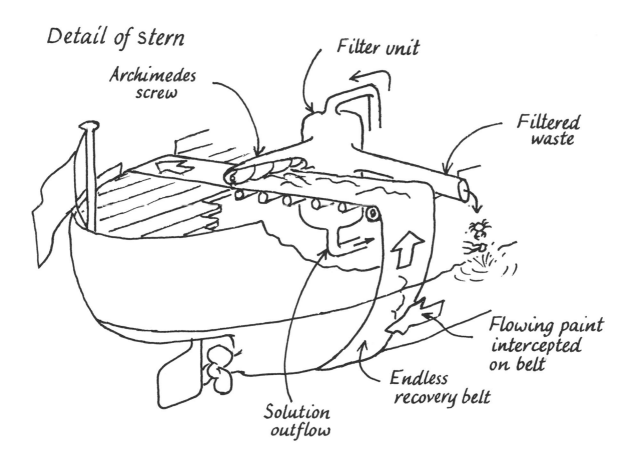

Filter unit

Archimedes screw

Filtered waste

Solution outflow

Endless recovery belt

Flowing paint intercepted on belt

The morality of eating meat

Rearing animals is a very inefficient way to obtain animal protein. Even a very efficient animal consumes about 3 pounds of food for every pound it gains in weight; and after slaughter much of the carcase is wasted or, at best, serves as fertilizer. There would seem to be a case for a system of battery-vultures to use up and recycle the unwanted portions and thus improve the overall efficiency of meat-farming. But Daedalus shrinks from inflicting such indignities on a noble bird, and instead is pursuing quite a novel way of obtaining meat. It is based on the observation that many kinds of small lizard, if seized by the tail, will escape by shedding the tail, and growing a new one later. He reckons that a proper programme of selective breeding should re-establish this reflex in the closely related but much larger crocodiles and iguanas; both species have thick tails containing a high proportion of meat. The animals would be kept on large farms where the crocodiles, as carnivores, would consume unwanted animal offal, while the iguanas would be given fodder. Each animal by repeated seizings would provide a steady succession of nutritious tails and might during its lifetime provide many times its own weight of meat, without the necessity of conventional butchery. Crocodile-skin for handbags would also be obtained in this ecologically irreproachable manner; indeed, creatures now facing extinction from man's thoughtless predation would be given domestic protection and their survival would be assured. The meat could well be commended to animal-lovers by slogans like 'From Contented Crocodiles!'

But this approach of regarding the meat as separate from the animal can be taken even further. The laboratory technique of tissue-culture enables cell masses and even recognizable body components to be kept alive in oxygenated nutrient media. So Daedalus sought for the biggest natural bone-free and fully-utilizable animal tissue which might be kept alive by such methods; he reckons it is the elephant's trunk. A fresh trunk might well be kept alive by circulation of appropriate media through its natural ducting, or else on a proper heart-and-lung machine. Furthermore, soft tissues grow in response to the stresses imposed on them; so a trunk kept in tension by a system of rollers would continue to grow indefinitely. Thus would be produced an endless supply of tasty trunk without the waste and cruelty of maintaining and culling large elephant-herds. Machinery which wrapped pastry continuously round the extruding trunkage and passed the product through a heated zone for cooking would complete a process, lending itself to full automation, for producing continuous elephant-trunk pie.

(*New Scientist*, 18 February 1965 and 16 March 1967)

Daedalus comments

Within a few years, both these schemes had reached the stage of serious advocacy, though not with the species I had suggested. In 1970, D.M. Skinner and D.E. Graham described in *Science* (Vol. 169, 6 August, p. 383) their discovery that the Bermuda land crab *Gecarcinus lateralis,* if it loses several limbs, goes into a precocious moult and regenerates them, with very little danger to its survival. They suggested that this habit in other crustacea could be economically significant: 'For example, the Alaskan king crab *Paralithodes camtschaticus . . .* is being depleted. Since walking legs are the source of meat in this species and since the animal regenerates limbs, it might be commercially desirable to remove 4–6 walking legs at the time of capture and return the animals to the sea. This procedure would leave the animal not incapacitated, could be predicted to stimulate precocious molting with attendant regeneration of walking legs, and would not deplete the population.'

And the idea of tissue-culture as a source of meat was advocated on both moral and economic grounds by D. Britz in a letter to *Nature* in 1971 (Vol. 229, 5 February, p. 435). In a subsequent correspondence in the journal, G.E. Moore of Roswell Park Memorial Institute, Buffalo NY (Vol. 230, 12 March 1971, p. 133) argued that it would be prohibitively expensive, while S.J. Pirt of Queen Elizabeth College, London (Vol. 231, 7 May 1971, p. 66) felt that it could be quite competitive. Both agreed that it was possible. In a rather chilling aside, G.E. Moore commented. 'We have fed residual cultured human cells to tropical fish for several years, and can testify that the diet was apparently nutritious, supported rapid reproduction, and was not associated with the development of tumours.' So purchasers of tropical fish in Buffalo, NY, beware! They may have acquired the taste for man-eating!

Walls have ears

Daedalus has been musing on the tantalizing aspects of dead languages, of how we can never tell how Latin or ancient Greek sounded from mere perusal of the written symbols. Little things give clues, like the phrase Aristophanes attributed to his frogs: 'Brekekekex Koax Koax' — always assuming Greek frogs have not extended their limited repertoire since his day. But what is really needed is some natural recording process by which ancient sound was authentically entrapped. Daedalus, impressed by the vocal artistry of his interior decorators, now believes that this is provided by the humble art of plastering. He points out that a trowel, like any flat plate, must vibrate in response to sound: thus, drawn over the wet surface by the singing plasterer, it must emboss a gramophone-type recording of his song in the plaster. Once the surface is dry, it might be played back by running a pick-up trowel over the surface in the same direction, though better fidelity could be achieved by first casting a more durable replica of the surface, or reconstructing the wave-form by microscopy.

So, here is a quite new and very powerful method of recovering the work-songs of ancient Greek plasterers. A rich vein of classical bawdy ought to be revealed in, e.g., the stucco of the palace of Knossos; respectable antiquarians given to voicing the hopeless plea, 'If only these walls could speak!' will probably be somewhat dismayed by their tone when they do. But much wider fields are open to Daedalus's new technique of archaeographony. Thus the dictation of letters taken down with the stylus on clay tablets must have left their vocal equivalent embossed into the tablet, together with informal expletives when the hapless scribe made a mistake. Daedalus hopes to examine the plaster of old Stratford on Avon and finally disprove the deplorable theory that Shakespeare spoke American, which language was transported by the Pilgrim Fathers to the States where it fossilized, while we progressed towards modern BBC.

(*New Scientist*, 6 February 1969)

Daedalus comments

A little while after this piece appeared, *New Scientist* received and published (24 April 1969, p.201) the following pained letter:

Coincidence

Sir, — This is one of those very, very odd coincidences, I am sure.

I refer to the *New Scientist* of 6 February, p. 308, where Daedalus . . . 'points out that a trowel, like any flat plate, must vibrate in response to sound: thus, drawn over the wet surface by the singing plasterer, it must emboss a gramophone-type recording of his song in the plaster. Once the surface is dry, it may be played back' (etc.) . . .

How very odd, that I should have sent to *Nature,* a paper (dated 13 January, 1969) entitled '*Acoustic Recordings from Antiquity*'; which paper was perfunctorily rejected as being 'too specialized'.

In my paper I noted my early experiments (1961) in the recording of sound (music, voices, etc.) on clay pots and on paint strokes applied to canvas (as in oil paintings) and the *successful* reproduction of such sound using a crystal cartridge phono pickup and a spatulate, wooden 'needle'.

I point out in my paper that adventitious acoustic recording in past times might be found in 'scratches, markings, engravings, grooves, chasings, etc. on or in "plastic" materials encompassing such as metal, wax, wood, bone, mud, paint, crystal and many others.'

With the developments in electronic signal analysis which more and more can ferret out 'signals' buried in 'noise', it seems to me that this whole field of acoustic archaeology warrants serious attention.

Richard G. Woodbridge, III

North Road RD-2
Princeton New Jersey 08540
USA

After working on a project for some years, and finally preparing an announcement for the scientific press, it must be a distressing experience to wake up one morning and find the whole thing expounded in the Daedalus column. I wrote back to Richard Woodbridge expressing regret for the coincidence, and pleading innocent to the implied charge of rifling *Nature*'s waste-paper basket. Dr Woodbridge finally published his paper in the *Proceedings of the IEEE* (Vol. 57(8), August 1969, p.1465). He gave examples of successful 'recordings' of music on paint-strokes, and of finding an actual word acoustically embossed on a brushstroke in an oil portrait. But mine remains the earlier publication!

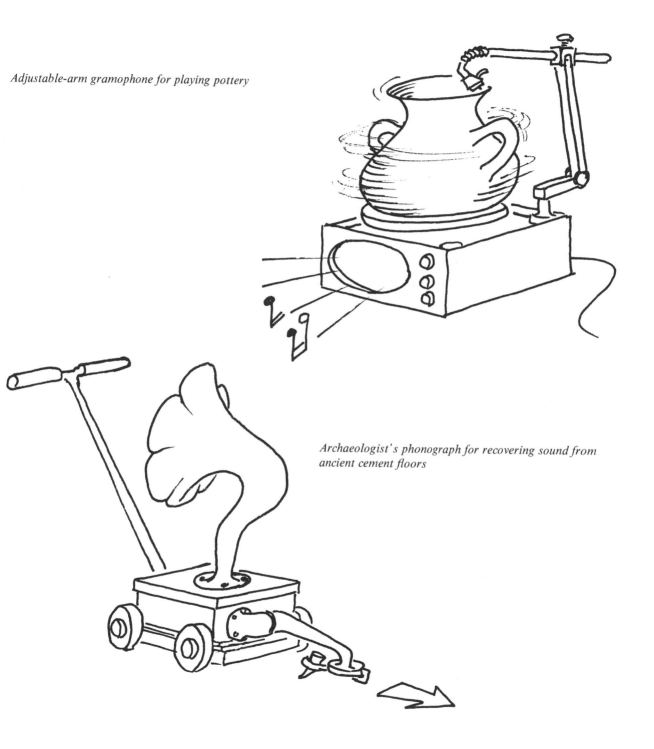

Adjustable-arm gramophone for playing pottery

Archaeologist's phonograph for recovering sound from ancient cement floors

The unisphere

The unicycle would be an outstandingly elegant and practical vehicle were it not for its lack of stability. At every instant, the rider must decide the direction in which he is toppling, and must then turn in that direction and pedal into the fall. Nowadays unstable systems are very easily servo-controlled — think of all those modern unstable warplanes that need a computer fighting full-time just to keep them straight and level — so Daedalus is devising a servo-stabilized unicycle, or rather unisphere. The rolling element is a sphere a foot or so across, the upper part of which fits into a cage equipped with motors and drivewheels. The rider sits on a saddle projecting up from this framework. Should he begin to topple, accelerometers detect the movement instantly, and the onboard microprocessor commands the motors to roll the sphere in the proper direction to frustrate the fall. With fast enough electronics the rider will feel absolutely safe, for the continual balancing tremors will be smaller than those which his own balancing reflexes would have to make were he simply standing up.

Daedalus's original idea was that the machine should be pedalled like a normal bicycle, and steered and braked by a simple tiller control; for like the bicycle, a unisphere would relieve the rider's legs from having to support his weight, freeing them to propel him at high speed and efficiency. A small battery charged by dynamo-offtake would supply the electronics and the balancing-motors, and could also provide power for short-term (e.g. uphill) boosts. But Daedalus then mused that a man on a unisphere would take up the same sort of space as a man on foot; and he now sees his invention as far more revolutionary than any mere improved bicycle. For a bicycle is a mode of transport. You pedal it where you want to go, then leave it locked to the railings while you go inside. But you never need get off a unisphere: you could use it everywhere, indoors as much as outside. Accordingly, it would have frequent access to mains electricity for re-charging, making an all-electric model feasible. This has several great advantages.

Firstly, the rider could navigate through doorways and in crowds without his pedalling feet and legs getting in everybody's way. Secondly, a really elegant, natural method of control and steering becomes possible. The rider merely leans in the direction he wants to go; the unisphere obediently rolls in that direction to counter the fall, and continues to roll that way until he leans to one side to change direction, or leans back to stop. With practice, this technique of balance-control — which is merely a generalization of the art of steering a bicycle, itself a simple extension of our automatic balancing reflexes — will become quite automatic. The rider will navigate his unisphere forwards, backwards and sideways without even thinking about it.

Thus the inconvenience and imperfections of our recently-evolved upright stance will be overcome, for the universal unisphere will take all the loads off our legs. People on unispheres will travel the highways and converse in corridors. They will stroll round museums without their aesthetic appreciation being sabotaged by aching feet; they will jostle round crowded bars without fear of being knocked off balance, no matter how drunk they or their companions may be. They will climb stairs without effort, and bring a new and mechanised eroticism to the dance-floor. Stools and chairs will become almost obsolete, and the disabled and arthritic will be able to lead full and active lives again. A whole new chapter will have to be added to the highway code!
(*New Scientist*, 18 May 1978)

From Daedalus's notebook

The main problem that the unisphere must surmount is climbing steps and stairs. It has to be given some sort of jumping ability. Since the saddle will have to be sprung anyway (e.g. by a telescopic-strut support) it should be possible to incorporate a linear motor into the strut. The total stroke of the assembly should be something like the maximum height of a step — say 30 cm. In normal travel, the saddle will be floated by the suspension-spring at about half-stroke. When the unisphere meets a step, the following sequence of actions must be triggered (either by the rider on his judgement, or under microprocessor control from the automatic step-detector on the front of the machine):

(a) The linear motor drives the saddle to full extension. If we restrict the acceleration of the rider to (say) $a = 0.5\,g = 5\,\mathrm{m\,s^{-2}}$, then over $l = 0.15$ m of travel the saddle will give him an upward velocity of $v = \sqrt{2al} = \sqrt{(2 \times 5 \times 0.15)} = 1.25\,\mathrm{m\,s^{-1}}$, and take a time $t = \sqrt{2l/a} = \sqrt{(2 \times 0.15/5)} = 0.25$ s. (If the rider weighs say 100 kg as an upper limit, the power required for this stroke is (force × distance/time): $P = (100 \times 5) \times 0.15/0.25 = 300$ W, which is not an excessive momentary load on the batteries.) The recoil of this stroke will compress the pneumatic sphere against the ground.

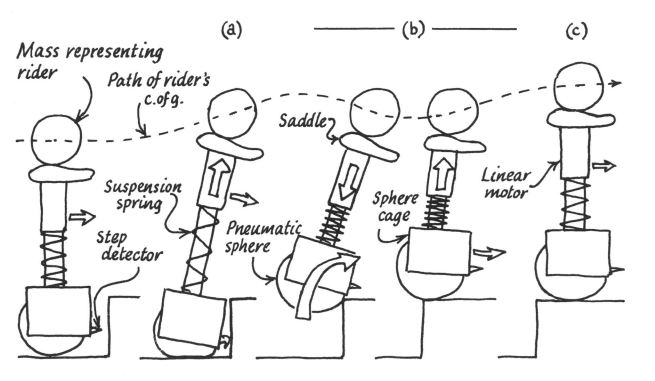

Mass representing rider

Path of rider's c.of g.

Saddle

Suspension spring

Step detector

Pneumatic sphere

Sphere cage

Linear motor

(b) The rider is now rising in 'free fall'. His centre of gravity will be travelling along a parabola whose horizontal component continues the motion of the unisphere before it reached the step, and whose vertical one will attain a maximum added height $h = v^2/2g = 1.25^2/(2 \times 10) = 0.075$ m and will then fall back again. During this symmetrical rise-and-fall its upward velocity will go from $+ 1.25$ m s^{-1} to $- 1.25$ m s^{-1}, taking a time $t = \Delta v/g = 2 \times 1.25/10 = 0.25$ s. For this period of time the unisphere is relieved of the rider's weight. The linear motor goes instantly into reverse, driving to the bottom of its stroke and compressing the suspension spring. This action, together with the released compression of the pneumatic sphere against the ground, will cause the unisphere to leap the step under the rider. To leap 30 cm demands an upward initial velocity of $v = \sqrt{2gh} = \sqrt{2 \times 10 \times 0.3} = 2.5$ m s^{-1} which doesn't seem excessive; and takes a time $t = v/g = 2.5/10 = 0.25$ s — just what we have available.

(c) The unisphere is now on the top step, having risen 30 cm. The linear motor is at the bottom of its stroke, so the saddle has risen only 15 cm — and so

has the rider. The motor now shuts off, so the compressed suspension-spring returns to its equilibrium half-stroke extension, lifting the rider the remaining 15 cm. For a single step (e.g. a kerb) this completes the jump-sequence; for a flight of stairs, it will be triggered again almost instantly by the next stair in the flight. The linear motor must then drive upwards from the bottom of its stroke to repeat the process. The jump-sequence should be held as a stored program in the balancing microprocessor, to be triggered whenever rider or step-sensor commands it. A reversed sequence, triggered by the 'free fall' of riding over a lip, could take the machine safely downstairs too.

Note. There's a minimum forward velocity needed to climb steps. The unisphere must move forward at least one pneumatic sphere-radius (say 15 cm) in the 0.5 s it takes to complete the jump, or it will be fatally unbalanced backwards on its arrival at the top of the step. So a forward speed of at least $v = l/t = 0.15/0.5 = 0.3$ m s^{-1} must be registered on the accelerometers, or attained by overbalancing against the step, before the jump-sequence can be initiated.

Raising the dust

No matter how diligently you dust the furniture, says Daedalus, more dust (or even the same dust) just settles back again. To solve this domestic dilemma, he began contemplating the aerodynamics of dust. Each dust-particle, typically an irregular fibre, will as it floats down adopt a stable attitude with its centre of gravity directly below its centre of drag. Now suppose you could spin it about this vertical axis. Unless it is utterly symmetrical, it must have some helicity about this axis and will act as an inefficient propeller. If its thrust is upwards it will climb. But if its thrust is downwards, this will reverse the aerodynamic forces on the particle. The centre of drag will become a centre of downward thrust, which will of course pull ahead to lead the centre of gravity. Thus the particle will turn upside down, and it too will climb.

Hence, says Daedalus, by spinning the dust in a room you could make it settle on the ceiling. He recommends a.c. electrodes in the wallpaper to set up a rotating electric field. This will induce a tiny dipole on each suspended dust-particle, which will then follow the field. So little energy is needed to spin a dust-particle that, feels Daedalus, quite a modest field of only a few hundred volts per metre should suffice. And the 50 Hz (3000 r.p.m.) spin-rate derived from normal mains frequency should give enough upward thrust to make the dust (which is tacky stuff) stick on the ceiling as it hits, or even to screw it in! So a room fitted with suitable embedded wall-electrodes wired to its mains supply would remain clean and sparklingly dust-free, while a useful sound-absorbing, heat-retaining blanket of dust would grow on its ceiling. For the obsessively house-proud, however, Daedalus is designing a helium-filled ceiling vacuum-cleaner.

The other use for this splendid principle is to keep snow off the motorways. Snow crystals are symmetrical, but the flakes themselves are irregular clusters of crystals which must have some propeller component. So as the snow comes down, rotating electric fields from electrodes in the crash-barriers could drive it up again. Only by drifting sideways from the road could it settle.

(*New Scientist*, 27 March 1980)

From Daedalus's notebook

There's a very interesting paper by N. B. Baravova and B. Ya. Zel'Dovich in *Chemical Physics Letters* (Vol. 57(3), 1978, p. 453). They point out that any molecule which doesn't have a plane of mirror symmetry must have some degree of helical 'handedness'. In other words it's a propeller; if rotated in solution it would experience a net hydrodynamic force along the axis of rotation. They want to use the principle to separate stereo-isomeric molecules, by spinning them apart in a rotating radio-frequency electric field. I'm not sure that it would work for molecules, but it should certainly work for larger objects, like dust-particles suspended in air. Very few dust-particles can have perfect mirror symmetry, so nearly all of them must be inefficient little propellers, and can be screwed along by a rotating electric field. The obvious application is domestic: a vertical-axis field would drive the dust-particles up or down (but note the effect of gravitational reversal on the downward-driven particles). A horizontal-axis field would screw the particles sideways so that they impacted the walls.

But there must be better uses for the principle than miserable dusting. Clearly a rotating-field electrostatic precipitator would fling dust out of industrial flue-gases with much greater efficiency than ordinary static-field ones. And as each new particle spins into the previously-deposited dust layer, its induced dipole and rapid rotation should screw it in, giving a really interpenetrating fibrous mat. In fact with any luck you'd wind up with something very like paper (which after all is only a lot of fibres laid together by filtering them from water suspension). So maybe novel abrasive, fireproof papers could be made by rotary electrostatic precipitation of power-station ash-bearing stack gases, cement-kiln fume, and so on. Come to that, how about a *real* carbon paper, made entirely of compacted soot? If the properties of carbon fibre are any guide, it might be extremely tough!

DREADCO's rotating-field midge deflector

But those unsmelt are sweeter

The springtime habits of frogs, says Daedalus, require some explanation. These engaging creatures often hop long distances towards water for mating purposes, and he wonders how they can locate it. The common idea that the animals can 'smell' water as such is obviously impossible, as their noses are, of necessity, permanently wet. Even worse, the permanent humidity of the atmosphere must swamp any tiny contribution from a distant pond. The fact that frogs patronize some ponds regularly every springtime but oddly neglect others, suggests to Daedalus that they sniff out a substance usually, but not always, present in pond water. By concentrating and analysing samples from the most popular frog-ponds, he hopes to isolate this water-indicating substance in pure form. Not only will it make possible an infallible frog-lure for disappointed amphibian-lovers and, possibly, hungry Frenchmen; it opens up a wide field of further research. For many other creatures appear to smell out water: and indeed the erratic but sometimes convincing performances of water-diviners suggests that they too operate in just this way. Diviners are seldom able to explain how they work; so Daedalus suspect they detect 'unconscious smells' which register without attracting attention, like the body effluvia which convey imperceptibly the collective atmosphere and emotion of an assembly.

These divining substances could be very useful. A pub releasing a concentrated diviner-compound could attract thirsty customers for miles around without them realizing why they were suddenly feeling so very thirsty. Ice-cream-van drivers, whose product also appeals to the sense of being hot and dry, could also establish their conditioned reflex among the masses without making the air hideous with metallic dissonance. Even public lavatories could, in some psychologically converse manner, attract desperate customers who now wander hopelessly around wondering where the nearest convenience is. But all this assumes that the effluvia affecting men are distinct from those affecting animals — otherwise publicans, ice-cream men and urinators alike will be engulfed by hordes of amorous frogs!

(*New Scientist*, 1 April 1971)

Dimly perceived smells can be very significant. We all know homes whose smell strikes the visitor so forcefully it seems impossible that the inhabitants fail to notice it; but clearly they only register it subconsciously as the 'home smell'. Daedalus reckons that like many animals, humans use such subconscious smells to identify safe surroundings, familiar people, etc. He attributes that uneasy 'settling-in' period in a new house to the slowness of uptake of the family smell by the new structure. So his smell-recorder passes domestic air ceaselessly through a liquid-nitrogen-cooled trap to condense out the odorous volatiles. Gradually it will accumulate enough of the house-odour to bottle in concentrated form as an aerosol. A quick spray-around will then instantly saturate a new house and make it seem like home again. Similarly multinational companies, who often go to great pains to establish a consistent 'house style', could also enforce a 'house smell'. They would use centrally compounded aerosols to enforce this uniform company odour in all their establishments. Then they could transfer managers all over the place at whim without suffering the long drop of efficiency as the uprooted executives re-acclimatize. On a smaller scale, the commercial traveller could take his home-aerosol on his travels, and spray hotel bedrooms, railway compartments, etc., with it. Instantly he would feel at home, and all the vague tension and unease of being on foreign ground would vanish. Perhaps he could similarly record and store a 'wife-smell' to spray over his transient girl-friends and thus remove the subliminal guilt of infidelity — but then maybe much of its allure would vanish at the same time. Certainly mothers could record their smell in aerosol form for baby-sitters to use in pacifying distraught offspring. And since some substances like tellurium impart a very long-lasting characteristic smell to the sweat of anyone who swallows them, a wonderful 'elixir of human brotherhood' might be possible. Put in food or the water, it would give the smell of the whole human race a common component, and thus subliminally cement international and interracial friendship!

(*New Scientist*, 5 January 1978)

'Ice-cream men . . . engulfed by hordes of amorous frogs!'

The optically flat Earth

The atmosphere thins out steadily with increasing height above the Earth's surface. An intriguing consequence of the resulting refractive-index gradient is that light traverses the atmosphere in a slight curve, tending to follow the curvature of the Earth. Thus you can see the Sun over the curve of the world 2 minutes after it has really set. Daedalus's calculations on this effect show it to be remarkably delicately balanced. Indeed, if the Earth had a radius only 13 km smaller, a ray at the surface would follow its curvature exactly, and it would appear flat. Departing ships would not sink below the horizon but merely dwindle into the distance; and people would not have realized the Earth was round until they discovered that, with a good telescope, you could see the back of your own head.

Daedalus calculates that this lost opportunity for extended vision can easily be recovered. Replacement of the surface air by a heavy blanket of sulphur dioxide would do it, this gas having exactly the right density and refractive index. But rather than lighting sulphur bonfires on the tops of mountains to see over them, he advocates constructing long pipes full of the gas, which would appear straight no matter how far over the curve of the world they extended. Down these visual corridors, with the aid of high-magnification gas-tight periscopes, people could smile cheerily at distant relatives and high-volume optical-image data could be transmitted. Once some really long transcontinental data-pipes were established, Daedalus would like to place mirrors at each end to experiment with precision time-of-flight spectroscopy (bouncing photons back and forth to time them). But capital is only likely to be forthcoming for pipes carrying gases more important than sulphur dioxide. Both ethylene and methane are carried long distances by pipeline these days, so Daedalus is trying to persuade the gas and chemical companies to lay their pipelines in long great-circle sections. He calculates that an ethylene line pressurized at 2.1 atmospheres, or a methane one at 5.9 atmospheres, would have just the right refractive index gradient to carry light as well.

(*New Scientist*, 19 October 1972)

From Daedalus's notebook

How does the refractive index n of a gas vary with height? Well, $(n - 1)$ is proportional to density, which falls off exponentially with altitude. So the index n_h at height h is given by:

$$(n_h - 1) = (n_0 - 1)\exp(-h/H)$$

where $H = RT/gm$ is the 'scale height' over which the density drops by a factor of e (m is the molar mass of the gas), and n_0 is the index for zero height ($h = 0$). So

$$n_h = 1 + (n_0 - 1)\exp(-h/H)$$

and differentiating,

$$dn_h/dh = \frac{-(n_0 - 1)\exp(-h/H)}{H}$$

Now for a beam to travel parallel to the Earth's curved surface, the upper ray must go proportionally faster over its longer journey, keeping the wave-front always vertical. Since speed is inversely related to refractive index, we require at radius $r =$

$$n_h r = n_{(h + dh)}(r + dh)$$

i.e.

$$-dn/dh = n_h/r$$

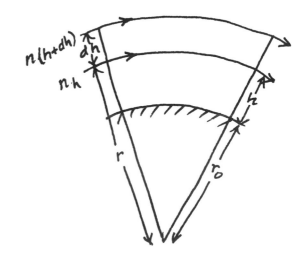

In the atmosphere this condition will be met at some radius $r = r_0 + h$, corresponding to some height h above the Earth's surface:

$$- dn/dh = n_h/(r_0 + h)$$

Replacing dn/dh and n_h by the expressions for them above we get:

$$\frac{(n_0 - 1)\exp(-h/H)}{H} = \frac{1 + (n_0 - 1)\exp(-h/H)}{(r_0 + h)}$$

which can be juggled into a simpler form:

$$\exp(h/H) = (n_0 - 1)[(r_0 + h)/H - 1]$$

Since both h and H are very small compared to the radius of the Earth, $r_0 = 6.37 \times 10^6$ m, this is practically equivalent to:

$$\exp(h/H) = (n_0 - 1)r_0/H$$

or

$$h = H \ln[(n_0 - 1)r_0/H]$$

So all we have to do now is look for a gas whose refractive index at one atmosphere, n_0, and scale height H gives h = zero. This gas will then bend light right round the Earth at the surface.

The best bet seems to be SO_2, for which at 20 °C $n_0 = 1.000\,615$ and $H = 3880$ m, giving $h = 37$ m only — quite negligible on this scale. The industrial gases ethylene and methane (n_0 at 20° C 1.000 648 and 1.000 411 respectively) have indices too low for their scale heights and give large negative values of h. Pressuring them to 2.1 and 5.9 atmospheres respectively, raising their $(n_0 - 1)$ values in proportion, just does the trick.

And what about air? At 20 °C it has $n_0 = 1.000\,272$ and $H = 8560$ m, giving $h = -13\,700$ m. So 13 km underground you could dig a tunnel curving with the Earth's surface and see along it to any distance. Alternatively, if the Earth were only 13 km smaller in radius, we could see round it!

Pulsed laser

Circum-global tunnel

13 km

Hinged mirror.
If it retracts within 0·13 sec after a pulse, the pulse will circulate in endless orbit

The last of the big suspenders

The seed-burrs of the teazle, which cling so tightly to fur and fabrics that brush against them, were the original inspiration behind the development of that similarly micro-hooked gripper-material 'Velcro'. Daedalus now reveals a related biological inspiration of his own. He recalls as a young lad dropping those ears of wild barley down his elders' tunics, so that they worked maddeningly around their victim's body by the ratchet-effect of their directional hairs. He now suggests that garments lined with a similar directional fur could be actively self-adjusting. This would free them from the almost inevitable requirement of being anchored at waist, shoulder or other natural body constriction. Even reputedly self-suspending stockings only stay up by passive clinging; by contrast, Daedalus's active socks and stockings would constantly seek to climb the leg as the wearer moved about, taking up any slack that might develop. Similarly the self-centring collar and tie would eliminate the distressing tendency of this appendage to settle under one ear (especially during interviews). In fact Daedalus envisages a fully automated clothing-outfit which, slung on anyhow, would actively ratchet its way to symmetrical and unwrinkled perfection not despite but actually because of its wearer's frantic activities. But Daedalus's new fabrics open up whole new fields of clinging and self-supporting garments, and he trembles to think of the bizarre creations now realizable by fashion designers. The sketchiest creations now no longer need discreet suspension systems, and even kinetic clothing (e.g. the endlessly rotating cummerbund) is possible. Daedalus prefers to consider saner applications of his principle: stair-carpets which climb the stairs under heavy traffic instead of slumping down, and office-chairs which maintain the sleepiest of individuals in a posture of alert eagerness.

(*New Scientist,* 17 August 1967)

Daedalus comments

At the time of Hayward Marum's enquiry, stockings were fighting their last battle against tights, so I suppose the industry was unusually receptive to possible innovations. But despite a detailed and encouraging reply to their letter, they seem not to have commercialized the idea.

From Daedalus's notebook

Self-climbing stocking

Directional ratchet-fibres

Self-adjusting tie (problem: may strangle wearer?)

Upwardly mobile stair carpet

Ratchet-fibres

HAYWARD MARUM INC.

122 S. MICHIGAN AVENUE, CHICAGO, ILLINOIS 60603

CORPORATE OFFICE

October 14, 1968

New Scientist
128 Long Acre
London WC 2
England

      ATTN: Editor of Ariadne Column

Dear Sir:

While reviewing an old issue of <u>New Scientist</u>, August 17, 1967, I read in Ariadne about an idea for "active socks and stockings (which) would by contrast constantly seek to climb the leg as the wearer moved about taking up any slack that might develop."

I would be interested in speaking to anyone who has some innovations of that nature. Looking forward to hearing from you.

Cordially,

HAYWARD MARUM, Inc.

Robert J. Calvin
President

cdh

Marum ∎ Christion Dior "socks for the man"
Sales offices: 40 East 34th Street, New York, N. Y. 10016
Plant: 15 Union Street, Lawrence, Mass. 01840

Hayward ∎ Thread-O-Life "hosiery for the lady"
Sales offices: 385 Fifth Avenue, New York, N. Y. 10016
Plant: Ipswich, Mass. 01938
A & L - Missy Hose
High Point, N. C. 27261

Spin, bonny boat

A solar-powered ship, says Daedalus, would be very underpowered even if it captured all the sunlight falling on it. He is accordingly designing one which exploits the sunlight which falls on the sea around the ship. Imagine, he says, a tall chimney, widening into a funnel at its base, and supported a few metres above the sea surface. Water-vapour evaporating from the sea would collect in the funnel and rise (because, with its low molecular weight, it is much lighter than air). This would suck in vapour-laden air from outside the chimney, and this would enter and rise in its turn. The faster the updraught the bigger the inflow around the base and the larger the surrounding area whose evaporation-energy would thus be captured as updraught in the chimney. The Coriolis effect of the rotating Earth would bring the inflow in spirally — in fact Daedalus has invented an 'anchored whirlwind'. Whirlwinds, like hurricanes and other circular storms, are indeed ultimately powered by sunlight in this sort of way, but in the absence of properly designed chimneys can maintain themselves only on a very large scale, and tend moreover to wander disconcertingly about. Daedalus's design, however, should entrain air from an area only a chimney-height in radius (say 100 m) and so intercepts only a modest 30 MW of sunlight. But even at a modest 3% efficiency, this should provide ample power for his 'tornado-boat'.

To convert the power to propulsive effort, Daedalus favours a daring design which exploits the rapid hurricane-like spin imposed on the twisting airstream as it converges on the vessel and rises in its chimney. Fixed vanes on the vessel and in the chimney will spin the whole 'rotor-tornado-boat' about its vertical axis. (This neatly gyrostabilizes an otherwise worryingly top-heavy design.) A system of cyclically varied angled vanes under the water will convert the spin to thrust in the desired direction, helicopter-fashion. Counter-rotated cabins will be provided for the crew. The craft will spin anticlockwise in the Northern Hemisphere, clockwise in the Southern Hemisphere, and will be momentarily becalmed as it crosses the Equator.

(*New Scientist*, 5 April 1979)

Daedalus comments

My original idea was for a sort of sea-going greenhouse. I envisaged a glazed dome floating on the sea like an enlarged cheese-cover entrapping air beneath it. Radiation entering would be absorbed through the greenhouse effect and the temperature inside would rise. If the sea outside were at $10\,^\circ$C (vapour-pressure 1230 N m^{-2}) while the water inside managed to reach $40\,^\circ$C (v.p. 7370 N m^{-2}), the excess internal pressure might provide propulsion as well as buoyancy. Since a solar pond with a dark radiation-absorbing bottom can heat the bottom water-layer almost to boiling, this seemed quite modest. Possibly it would help to pour cuttlefish ink onto the internal water-surface to make sure that all the radiation was absorbed there. The excess internal pressure could propel the boat by jet reaction merely by opening horizontal vents in one side of the roof. This is a very neat solar boat with no moving parts and little wetted surface, but there are several drawbacks. Firstly, unless the size of the vent openings is continuously and carefully matched to the rate of evaporation within, the craft will fill up and sink; secondly, even if very wide and squat it would probably be unstable against capsizing, and would need stabilizing floats round the edge; thirdly, efficiency is bound to be low. A 100-m square version might at the very outside capture 10^4 kW of sunlight; but such a low-pressure jet engine working from such a small temperature difference could hardly reach 1% efficiency. That implies at most 100 kW of useful power, which is not much to push a 100-m-square greenhouse around. Clearly a feasible solar ship has to capture the sunlight from a far wider area of sea than it can cover itself.

On this philosophy a sailing ship is worth imitating. It uses the wind created by solar convection over millions of square kilometres of sea. Could I induce a purely private, local wind for my vessel to exploit? In a very thoughtful article, J. D. Bernal (*The Scientist Speculates*, ed. I. J. Good, Heinemann, 1961, p. 17) points out how fundamental an invention the chimney was, and how rapidly moist air would rise in one. This is not because it is hot but because it is inherently light: the molecular weight of water-vapour (H_2O = 18) is considerably less than that of air (average m.w. = 29). So a chimney sucking in and channelling the moist air evaporated from an area of sea, would create and concentrate a 'local wind' by defining a local convection cell. The obvious way of using the up-draught-energy was via the rotation of fans in the chimney. This led me to recall how a tornado amplifies the slow Coriolis-rotation of the atmosphere by sucking in air from a wide area and reducing its radius of rotation. The conservation of angular momentum gives the air a rapidly increasing rotation as it nears the centre. The final design followed fairly logically.

THE ROTORTORNADOBOAT

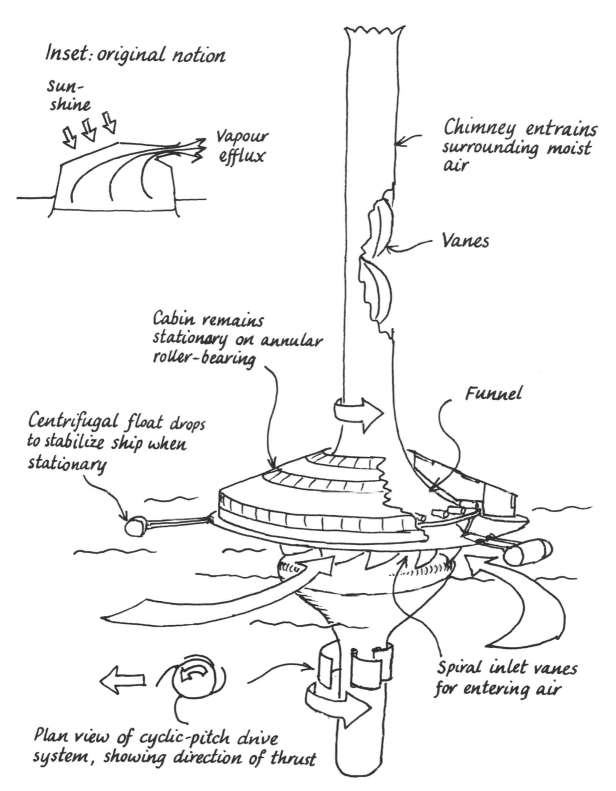

Inset: original notion

sun-
shine

Vapour
efflux

Chimney entrains
surrounding moist
air

Vanes

Cabin remains
stationary on annular
roller-bearing

Funnel

Centrifugal float drops
to stabilize ship when
stationary

Spiral inlet vanes
for entering air

Plan view of cyclic-pitch drive
system, showing direction of thrust

Magnetic monopoles

Like many physicists, Daedalus has been exercised by the lack of symmetry between magnetic poles and electric charges. The electromagnetic equations are quite symmetrical; nonetheless, while you can easily isolate positive and negative electric charges, isolated north or south magnetic monopoles seem not to exist. (Anyone who tries to obtain them by sawing a magnet in half merely winds up with two smaller magnets.) Daedalus's approach to the problem is to consider what isolated monopoles would do if left to themselves. They would, of course, set off north or south along the lines of the Earth's magnetic field, and bury themselves at the Earth's magnetic poles. So Daedalus is trying to get support for a geological expedition equipped with sno-cats and drilling rigs to search for these caches of single poles. When found they would be ideal for making d.c. motors, for they would rotate endlessly around any current-carrying wire. But you would have to be careful not to let them go.

(*New Scientist*, 3 December 1964)

Daedalus comments

About four years after this piece appeared, I was amused to come across an article in *Science Journal* (September 1968, p. 60) in which Dr H. H. Kolm, of the MIT Francis Bitter Magnet Laboratory, described his geological searches for magnetic monopoles.
So far, he hasn't found any.

The possible natural occurrence of magnetic monopoles continues to tantalize the scientific community; and Daedalus, who has a lot of uses lined up for them, now outlines a simple monopole synthesis. Consider a spherical steel ball cut into sectors, and each sector magnetized so that the surface-end is north and the centre-end south. Then, he argues, when the sectors are reassembled the central south pole will be completely shielded by the surface north one; effectively a spherical north monopole will result! If released, these novel spheres will head north unless trapped into endless rotation around current-carrying wires (forming the simplest possible motor).

At first Daedalus planned to exploit the mutual repulsion of monopoles in vehicle-suspensions, by assembling appropriately magnetized sectors of steel cylinder into two long parallel monopoles. A repelled spherical monopole would then levitate stably between them, and they could be joined up at the track ends to prevent leakage of flux. But he then realized that free monopoles, repelling each other, will distribute themselves gas-like throughout a volume, exerting pressure on its walls. So a much simpler hovercraft could be supported by a 'magnetic gas' of monopoles underneath, retained by a copper gauze skirt. Not only will this perfect suspension be free of the power-consumption and noise of the conventional hovercraft; it will have a steady thrust north. So a magnetic hovercraft track, if kept free of d.c. power cables which might trap the vehicles into endless orbit, could serve the key north–south corridors with silent, fumeless, smooth-running vehicles propelled by their suspension monopoles! Ample supplies of the appropriate class of monopoles would be stocked at either end, and the energy for the system would come from the locomotives hauling bulk monopoles under pressure back against the Earth's field for recycling. But difficulties might arise where trains carrying monopoles of opposite polarity passed on their respective journeys ...

(*New Scientist,* 10 September 1970)

Daedalus comments

This monopole-synthesis has since been analysed rather unsympathetically by Dr Epsilon and by Mr J. Middlehurst (*Wireless World,* December 1978, p.67; and September 1979, p.82).

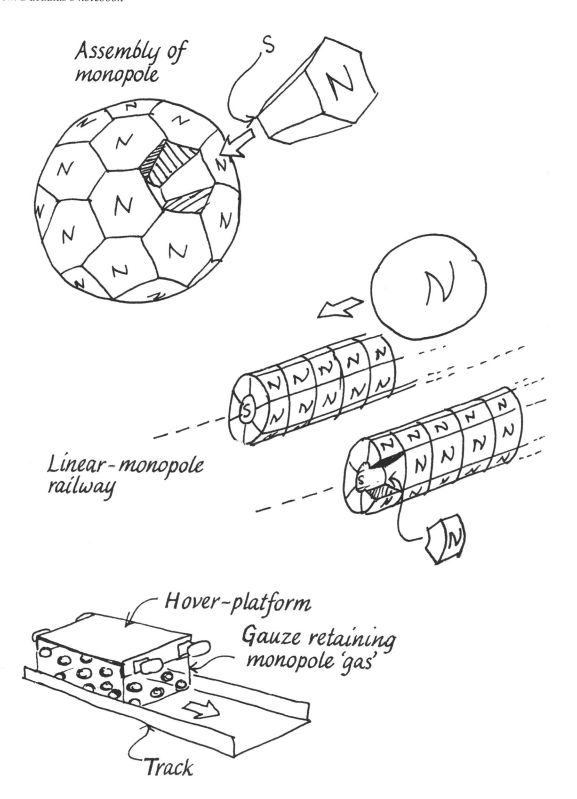

Assembly of monopole

Linear-monopole railway

Hover-platform

Gauze retaining monopole 'gas'

Track

Antidotes to wear

The tungsten–iodine lamp is one of the cleverest of chemical inventions. As in all incandescent lamps, the filament slowly evaporates away; but the resulting metallic vapour soon reacts with the iodine in the lamp. The tungsten iodide vapour thus produced is decomposed by white heat, so when its molecules encounter the filament they deposit the tungsten back again! Daedalus sees in this brilliant device a whole new philosophy of self-repairing mechanisms. He has long deplored the absurdity that condemns a 1-ton car to the scrap-heap because of a gram of metal worn from the moving parts, and has been seeking to apply the tungsten–iodine principle to eliminating wear in bearings. He recalls the Mond nickel process, which exploits the reaction between nickel and carbon monoxide to volatile nickel carbonyl. At a higher temperature this carbonyl decomposes again, depositing the extracted nickel and freeing carbon monoxide for recycling. So he is modifying his own long-suffering jalopy by nickel-plating the cylinders and bearings, and bubbling the exhaust-gases through the oil in the sump. The idea is that wear in the moving parts will remove nickel from them and deposit it as fine particles in the sump-oil, where it will react with carbon monoxide from the exhaust. The nickel carbonyl thus liberated in the oil will circulate with it through all the bearings and cylinder surfaces. Now the hottest spots the oil will encounter are just those where friction is causing wear — so the thermally decomposing carbonyl will replace the metal exactly where it came from and the car will run for ever!

Clearly this principle strikes a fatal blow at the shoddy philosophy of planned obsolescence, and designers will hesitate before perpetrating grotesque outrages destined to endure down the ages. Almost all engineering objects could benefit from wear-free mechanism: cars, aircraft, instruments, machine-tools that must maintain their precision in very abrasive environments, even sewing machines and micrometers. But care would be needed to maintain the chemical balance of the system. If too much metal were deposited on a hot-spot, its clearance would not merely be restored, but reduced. It would run tighter and hotter, attract even more metallic deposit, and soon seize solid. So just enough dissolved metal should be present in the system to keep all bearings at their proper clearances, and over-lubrication should be avoided.

(*New Scientist*, 14 December 1967)

Daedalus comments

Within four years the mechanism I proposed here for a self-healing bearing was discovered by lubrication scientists. One well-known design of bearing has a thin film of soft metal on one of its faces. This gives very desirable running characteristics but the soft metal film is steadily worn away. In 1971 J. D. Dickert, jun., and C. N. Rowe of the Mobil Research and Development Corporation reported (in *Nature Physical Science*, Vol. 231, 1971, p. 87) some experiments on the effects of certain additives on hexadecane lubricant in a standard pin-and-disc testing machine. One such additive was gold 0,0-di(neopentyl) phosphorodithioate. It appeared to decompose on the hot spots of greatest frictional drag, and to deposit a thin film of soft metallic gold. This of course lowered the friction dramatically. As they commented: 'The gold generally deposits preferentially on the surface in constant contact, which is the one having the sustained high temperature — the pin or the stationary balls. In effect the system becomes one of steel sliding against gold, and any gold lost due to wear is replenished by further deposition. Thus, the system achieves a balance of film wear and film formation depending on the severity of operating conditions, and with proper balance the bearing surfaces can survive for as long as additive is present in solution.'

Gold, of course, is rather an expensive additive to motor oil. But the principle has been clearly established. I wonder if Mobil engineers have some cheaper, maybe lead-based oil additive up their sleeves, or whether the next move is up to DREADCO? One field in which chemistry of this kind could be literally a life-saver is that of implant surgery. Artificial hip-joints, etc., can be amazingly successful prostheses, but their life is limited to 5–20 years because of wear and abrasion of their bearing surfaces.

Operation of tungsten-iodine lamp

(a) Tungsten atoms evaporate from hot filament

(b) Atoms combine with iodine molecules forming tungsten iodide

(c) Tungsten iodide molecules diffuse throughout bulb

(d) Tungsten iodide molecules hit the hot filament and decompose, redepositing tungsten

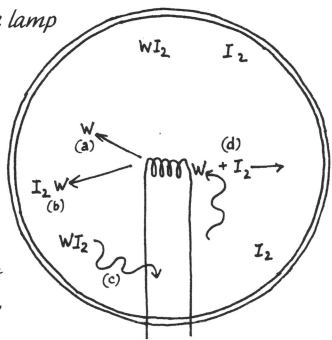

Nickel carbonyl anti-wear bearing system

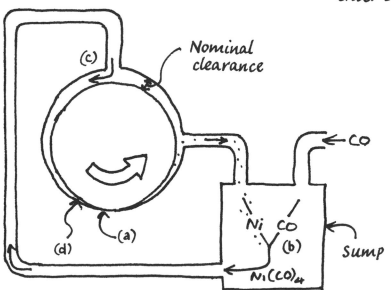

(a) Nickel particles abrade from loaded area of bearing, and enter oil stream

(b) Nickel reacts with carbon monoxide in sump, forming nickel carbonyl

(c) Dissolved nickel carbonyl circulates in the oil

(d) Nickel carbonyl decomposes on the hot-spots of the bearing, redepositing metal on the loaded area

Your tiny mind is frozen

Various optimistic souls nowadays are getting themselves frozen down in liquid nitrogen, in the hope of being revived in better days to come. Daedalus is sorry to point out that, at the moment, this cannot possibly work, because rapid and uniform freezing and rewarming are required. Now while it is possible to warm up a body quickly and uniformly by either dielectric or microwave heating (which penetrate evenly), current 'cold bath' cooling methods remove heat only through the skin. The frozen zone moving slowly inwards must create the same type of havoc as trying to stop a complex and interlinked machine a piece at a time. What is needed is some inverse process persuading the body to give up its heat as microwaves. Daedalus now points out that this is the essential of maser action, and proposes to achieve it via that well-known chemical tool, proton magnetic resonance (p.m.r.). Protons (hydrogen nuclei — the commonest nuclei in the body) tend to align themselves with an applied magnetic field. The alternative alignment against the field requires higher energy: in fact for strong enough fields the energy-difference between these two states corresponds to that of microwave radiation. Now, says Daedalus, suppose the field is suddenly reversed. All of a sudden, the protons will find themselves aligned against the field. The tiniest microwave stimulation will then trigger their descent to the stable alignment, with accompanying maser-emission of microwaves. By spin-lattice relaxation, the energy thus lost will be subtracted from the heat of the sample; and the process can then be repeated.

Daedalus has not yet called for volunteers for combined maser, refrigerator, and radar-emission duties. Instead his pilot studies use earthworms (just the right size and shape for the sample tube of his p.m.r. spectrometer). He is training them to tie themselves in a knot a fixed time-interval after an electrical stimulus. When they are proficient at this he will freeze them down bang in the middle of this interval. The crucial test will come when they are revived by microwave heating. They should complete the interrupted time-lapse and then obediently get knotted.

(*New Scientist*, 14 November 1968)

Progress has been maintained on the cold front initiated last week in DREADCO's laboratories. Daedalus's magnetic maser for instant people-freezing recalled those enigmatic quick-frozen mammoths of Siberia, some of which still have half-chewed greenery in their mouths.

Daedalus sees this as a natural instance of his rapid-freezing mechanism, presumably occurring as a result of one of the reversals that the Earth's field is known to have made during the Ice Ages, coupled with solar microwave irradiation via the disturbed ionosphere. Daedalus reckons that such rapidly frozen creatures must still be in suspended animation, and is now negotiating to send a party to Siberia with powerful microwave-heating gear to pulse them rapidly back to life. What a challenge to biology and animal psychology such a prehistoric animal would be, with its full set of stone-age reflexes! But it may not be so simple. Information can be stored in a brain in solid, material form: 'hardware' like interconnections between cells, chemical substances representing specific memories, etc. But it can also be stored as 'software': recirculating groups of nerve-impulses, electric charge patterns, and so on. The brain's hardware should survive freezing intact. But its software will almost certainly all be lost, just as the data in an electronic memory are lost on switching off. Daedalus is confident that vital routines like breathing, heartbeat and so on must be hardware in any animal; but reflex and memory may not be. If they *are* hardware, the resuscitated mammoth will recognize the party as dangerous humanity, and attack it in a rage. But if they are software the mammoth will recall nothing of its previous life. With the total naivety of the new-born it will then identify itself with the surrounding creatures and accept itself as human. Thus the experiment has powerful implications for people-freezing. Daedalus suspects that the versatile human brain is nearly all software, so the eventual defreezee will not recall why he opted out in the first place. Perhaps the maser-fridge has most potentiality as a latter-day equivalent of the Foreign Legion, for those who wish to forget.

(*New Scientist*, 21 November 1968)

Daedalus comments

The first frozen mammoths were found in Siberia in the eighteenth century, and a historical account is given by Pfizenmayer in *Les Mammouths de Sibérie* (Paris, 1939). More recent radiocarbon studies of Siberian mammoths (*Radiocarbon*, Vol. 4, 1962, p.178) indicate that the creatures have been frozen for over 39 000 years. A new and well-preserved specimen was discovered in 1971, and was transported, still frozen, to the Yakutsk Institute Permafrost Laboratory. No reports of its attempted resuscitation have yet reached the West, however.

Mammoth-resuscitation in Siberia

Radioastrology

Perplexed by the refusal of astrology to go away, Daedalus has been theorizing about it. One astronomical phenomenon that might possibly influence human affairs is the cosmic microwave background, which delivers its radiation to everybody all the time. The biological effects of low-intensity microwave radiation are still hotly disputed, so Daedalus feels that even this tiny dosage may influence the most sensitive tissue. And the most sensitive tissue is a freshly-fertilized human ovum, which by its thousand-millionfold growth in the womb is the most powerful amplifier imaginable for small initial perturbations. Furthermore, the Earth in its yearly travel around the Sun sees the microwave background with a changing Doppler shift — a maximum in November and a minimum in May. So the astrologers' tradition correlations are just 9 months wrong. It is the time of conception, not that of birth, which matters.

People conceived in November (therefore usually born in August) carry the impact of maximum-intensity microwave background; those conceived in May and born in February bear the minimum effect. And as in conventional astrology, the planets (which except for maybe Jupiter and the Moon, have a relatively feeble microwave intensity at the Earth) may contribute small second-order effects. But Daedalus goes further. Not only does the Earth go round the Sun; the Sun and Solar System go round the Galaxy, once per 200 million years. The cosmic microwave Doppler shift for this rotation is at least 10 times stronger, and should show up as a periodicity in the fossil record. Life is booming today (having just invented us). Half a galactic revolution ago, with the Doppler shift reversed, things were looking down: dinosaurs and ammonites were just dying out. But one revolution ago these creatures were just entering their boom phase. Two revolutions ago, in another mighty boom, life was just colonizing the land; while just three galactic revolutions ago the fossil record itself began. So, says Daedalus, microwave palaeoastrology is worth looking into. Be prepared for a slump in 100 million years time! Meanwhile, Daedalus is correlating the temperaments of laboratory gerbils with the microwave environment of their parents, and inventing microwave corsets for volunteer mothers-to-be.

(*New Scientist*, 31 May 1979)

From Daedalus's notebook

What physical effect might bias the personality of people in an annual cycle according to the month of their birth? It can't be climatic summer/winter alternations as (1) they'd vary with latitude and (2) they'd be reversed in the Southern Hemisphere, and astrologers say that only the time of birth matters, not the place. Other factors — e.g. the academic year which starts in the autumn, so a child born in September starts school earlier than one born in May — would also show a 6-month phase-shift in Australia.

One possible candidate is the cosmic microwave background, which is highly uniform over the whole sky and perfectly averaged out anyway by the Earth's daily rotation. However, it varies throughout the year because of the changing Doppler shift due to the Earth's travel around the Sun. R.A. Muller, among others, has measured the small variations in the microwave background (he has a good account of the matter in *Scientific American,* May 1978, p. 64), and it seems to be slightly 'hotter' in the constellation Leo and 'colder' in Aquarius.

Now the astrological month-signs denote the constellation occupied by the Sun as seen from the Earth. The Sun is centrally in Leo around 5 August, so (Diag.) the Earth is directly approaching Leo around 5 November. It therefore gains more microwave energy by the Doppler effect of its approach to the 'hot' direction. Conversely, it will receive least microwave energy around 5 May when it is receding from the 'hot' direction and approaching the 'cold' one. It makes more sense to correlate such variations with conception and rapid foetal growth, when small effects could be amplified, than with birth when, presumably, the characteristics of the child are pretty settled. But there seems no obvious way of testing that.

One thing we can test is the corresponding effect due to galactic rotation. The Earth goes round the Sun once a year at a velocity of $3 \times 10^4\,\mathrm{m\,s^{-1}}$. It and the Sun rotate once around the galactic centre every 200 million years, with a velocity of $3 \times 10^5\,\mathrm{m\,s^{-1}}$. This implies 10 times the Doppler shift against the microwave background, so the 200 m.y. galactic 'year' should have 10 times the astrological power of the solar year. Rather surprisingly, this fits the geological evidence quite well.

Taking the present as a time of booming and abundant life, the end of the Cretaceous 100 m.y. ago looks relatively depressing with the extinction of dinosaurs, ammonites, many brachiopods and many cycads. About 200 m.y. ago in the middle of the Triassic, these creatures were by contrast getting well into their stride. The Carboniferous period 300 m.y. ago doesn't look too bad, I fear, but the Devonian/Silurian period 400 m.y. ago was a great triumph, with the first land plants and animals appearing. So there may be something in it! Another line of the same evidence is suggested by Dr G. Paltridge's recent article on climate prediction (*New Scientist*, 19 April 1979, p. 194). His climatic-periodicity spectrum shows a peak at about 200 m.y., second in intensity only to the season-peak at 1 year.

Daedalus comments

H.J. Eysenck (*Encounter,* December 1979, p. 85) has discussed grounds for admitting the existence of 'cosmobiological' effects, which may or may not underlie the traditional claims of astrology. The microwave background hypothesis is a possible instance of such an effect. Eysenck notes the work of J.H. Nelson, a radio engineer with RCA, in correlating radio disturbances with planetary positions. I am tempted to cite this work as a way of hauling the traditional astrological planetary influences into the microwave hypothesis.

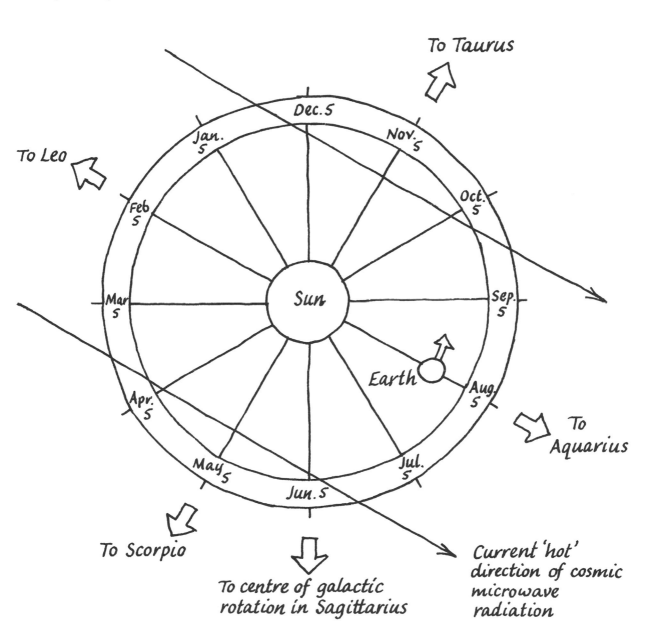

The UFO unmasked

Public interest in Unidentified Flying Objects, stimulated by space-fantasy films and many reports of actual sightings, remains high. So Daedalus has decided to come clean about these DREADCO prototypes (for such they are). They are sun-powered flying machines, rather like the Hughes Aircraft model aeroplane which powers its electric propeller from silicon solar cells (*New Scientist*, 9 March 1978, p. 659). The circular shape was originally chosen to pack the greatest radiation-collecting area into a compact shape. But it then seemed natural to give the design many small aerofins and generate lift by spinning the whole saucer, on a combination of helicopter and Frisbee principles. To prevent the pilot getting dizzy, or even being centrifuged flat at high revs, the central control dome forms a stationary hub.

This design has an intriguing and novel feature: it's an energy-storing flywheel. For in Britain's cloudy climate, any purely sun-powered machine is at the mercy of the weather for its take-off power. Only above 10 km altitude or so, above the clouds, can it be sure of sunlight. But Daedalus calculates that a saucer of light, strong carbon-fibre composite can easily be spun so fast that its rotational energy alone will suffice to lift it hundreds of kilometres. So DREADCO's flying saucers use spin as an energy bank. Revved up on the ground, they soar on their release to operating altitude, like that popular plastic helicopter-disc toy. Once above the clouds they acquire solar power, and can then govern their spin and direction of travel by feathering their aerofins as a helicopter does. Should they wish to drop below cloud-base they can use their fins in windmill fashion, to build up during their fall the added revs that will get them safely back up again. (These drops below the clouds are naturally kept short by the test-pilots, explaining the tantalizingly brief sightings characteristic of UFOs.) Finally, however, when they return to base, they build up to the original spin in their fall, returning to the launch-rotor the energy they borrowed from it in take off — an economy possible with no other type of plane!

(*New Scientist*, 4 May 1978)

From Daedalus's notebook

Is sunlight intense enough to lift a flying saucer? If the saucer is 5 m in radius, it will have $A = \pi r^2 = 75 \mathrm{m}^2$ of surface. So in full sunlight of about $1\,\mathrm{kW\,m}^{-2}$ and with 10% efficient solar cells it will generate electrical power

$W = 75 \times 1 \times 0.1 = 7.5\,\mathrm{kW}$. This power could lift a saucer of mass $m = 250$ kg (say) at a velocity $v = W/mg = 7500/(250 \times 10) = 3\,\mathrm{m\,s}^{-1}$. So to fly, the saucer must have a natural sinking speed in air of $3\,\mathrm{m\,s}^{-1}$ or less. This looks quite hopeful. Good man-carrying gliders sink at only $1\,\mathrm{m\,s}^{-1}$, and Frisbees sink uncannily slowly: they are aerofoils and gyrostabilized parachutes combined.

What energy can we store in a spinning saucer? The standard formula for the maximum spin one can give to a hoop before it bursts under the centrifugal stress is:

$$n^2 = S/(4\pi^2 r^2 \rho)$$

where n is the maximum spin-rate in revs a second, S is the tensile strength of the hoop, r is its radius and ρ its density. The linear rim-speed will be $v = 2\pi rn$, so the maximum kinetic energy E that can be stored by a spinning hoop of mass M is:

$$
\begin{aligned}
E &= \tfrac{1}{2}Mv^2 \\
&= \tfrac{1}{2}M(2\pi rn)^2 \\
&= \tfrac{1}{2}M(2\pi r)^2 S/(4\pi^2 r^2 \rho) \\
&= MS/2\rho
\end{aligned}
$$

This energy can lift the hoop to a height h given by:

$$
\begin{aligned}
h &= E/Mg \\
&= MS/2\rho Mg \\
&= S/2\rho g
\end{aligned}
$$

We want the strongest and lightest material for such a spin-stressed hoop: carbon-fibre reinforced resin. This has $\rho = 2300\,\mathrm{kg\,m}^{-3}$ and $S = 1.5 \times 10^{10}\,\mathrm{N\,m}^{-2}$ or so. So

$$
\begin{aligned}
h &= 1.5 \times 10^{10}/(2 \times 2300 \times 10) \\
&= 3.3 \times 10^5\,\mathrm{m} = 330\,\mathrm{km}!
\end{aligned}
$$

This is very encouraging. We only need to lift our saucer about 10 km from an initial launching spin, so we have a factor of 30 in hand to cope with inefficiencies, unspun weight, etc. Indeed, much of the energy needed for the whole flight might come from the initial spin, leaving the solar cells merely as a secondary topping-up system (It also looks as if we might be able to lift a carbon-fibre hula hoop at maximum revs right into orbit, but I'll not pursue that train of thought now.)

Mechanical details. Conservation of angular momentum, and cancellation of gyro effects, demand a contra-rotating design. The neatest layout has the top of the saucer as the 'stator' of a disc-pattern induction motor, spinning one way, the bottom half as the 'rotor', spinning the other way, with a differential gear between them providing a static spindle on which to mount the cabin. (The power from the solar cells will have to be inverted to a.c. to drive the induction motor, but that's no great problem.) The 'stator' can form a Frisbee-disc with circumferential entrainment-louvres for lift and stability, while the 'rotor' carries stubby aerofoil blades with cyclic- and variable-pitch mountings, helicopter-style, for control.

These blades would also enable the craft to be spun-up and steered in unpowered 'windmilling' descent — an important consideration in a rather weather-dependent vehicle with limited reserves of energy. But since to work at all the saucer can't have a natural sinking rate greater than a leisurely 3 m s^{-1}, unplanned descent shouldn't be dangerous. In effect, the craft is its own parachute.

Airflow through saucer

A. Air entrained by louvres for hovercraft / Frisbee effect

B. Air passing through aerofins for helicopter lift

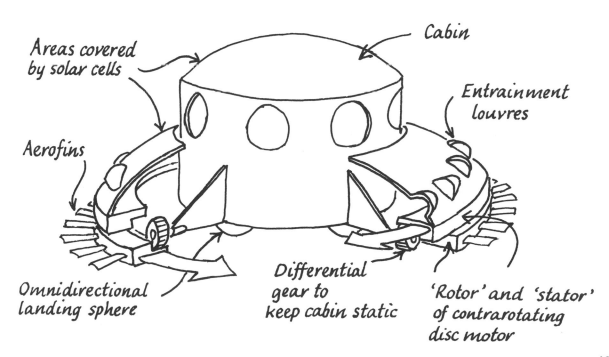

Cabin

Areas covered by solar cells

Entrainment louvres

Aerofins

Omnidirectional landing sphere

Differential gear to keep cabin static

'Rotor' and 'stator' of contrarotating disc motor

The soul and the photon

The soul and its reincarnation may seem a purely theological concern, but Daedalus has been looking at it scientifically. With (say) 10^{10} people in the world, a soul must carry at least 33 bits of information to be uniquely labelled. On the generous assumption that every living creature in the Universe since the Big Bang has had a unique individual soul, the figure rises to maybe 220 bits; repeated reincarnation of the same set of souls cuts it down again. Anyway, to impress 220 bits of information onto a chemical system at ambient temperatures needs at least 10^{-18} joules of energy, weighing some 10^{-32} grams. It would need a remarkably sensitive balance to detect this change in weight of a woman as she becomes pregnant, even discounting the embarrassing circumstances of the experiment and the difficulties of weighing two people in vigorous action. Fortunately theologians have decided that the soul enters the foetus not at conception, but at the moment the foetus acquires individuality. This is the point when the foetus, if divided, will no longer develop into twins, and occurs at about the sixth week of pregnancy.

Now, says Daedalus, the entry of spiritual information into the foetus must decrease its entropy. By standard thermodynamics this implies a decrease in heat capacity and the evolution of some 'latent heat of incarnation'. Assuming incarnation takes place instantaneously, as the indivisible, quantum nature of the soul would suggest, this heat energy should appear as a single photon. This makes the chances of detecting incarnation much better, as a 10^{-18} J photon is conveniently ultraviolet. So recently-pregnant DREADCO volunteers are wearing special photomultiplier corsets, aligned to detect this single photon and measure its frequency, from which its energy and information-equivalence can be deduced. Success will validate the new science of spiritual thermodynamics, and enable an upper limit to be calculated for the number of souls in the world. Should this number come out less than the total world population, Daedalus's base suspicions that there are a lot of zombies around will be confirmed.

This fascinating connection between the soul and the photon has a further implication. If the entry of a soul into a recently fertilized foetus causes emission of a photon, then death, the release of the soul, should reverse the process. It should be accompanied by the *absorption* of a photon by whatever part of the brain is entrusted with the ego. This momentary increase in UV absorption by a fairly accessible region of tissue should, Daedalus feels, be detectable by delicate spectroscopy. The ethical implications of inviting depressed DREADCO volunteers to commit suicide in an UV spectrometer are worrying, so animal experiments will be conducted first. If the increased UV-absorption on a creature's death tends to coincide with UV-photon emission from some pregnant female, then reincarnation of the soul would be strongly indicated. Studies could then be devised to show whether the soul prefers to enter a nearby female (an inverse-square law might be expected), or to enter one very like its previous mother. If the experiment succeeds, Daedalus will advocate equipping all recently pregnant women with his photomultiplier corsets. A photon-emission coinciding with a death elsewhere in the world will indicate a reincarnation. This might enable, e.g., a foetus reincarnated from an expiring genius to be identified for special education. But since the world population is growing, even total reincarnation must require a steady input of new souls, which must exist in disembodied form as a sort of spiritual field. Daedalus is already working on a theory of quantum theology, with a spiritual field whose individual 'soulon' particles play the part which virtual photons play in the electromagnetic field. Incarnation and death are then resonance-absorption and resonance-emission of soulons. If the resonances can be shifted by the Doppler effect, the probability of death or rebirth could be reduced by very high speeds — suggesting the use of Concorde as an emergency hospital, or defertilization clinic for unwilling mothers-to-be. But the increasing capture of soulons by an increasing population should in any case weaken the spiritual field, automatically reducing the chance of further soulon-capture and thus limiting fertility without sinful contraception. DREADCO is applying for a huge Vatican grant for this study.

(*New Scientist*, 17 and 24 November 1977)

From Daedalus's notebook

The purest theory of reincarnation holds that the departing soul enters a new creature with no knowledge whatever of its previous state. So the only information it can carry is the minimum 'label' needed to give it individuality — for all souls must be distinct. For 2^n different souls, these labels must map the 2^n binary numbers, each carrying n bits of information. So for a human world population of (say) 10^{10} souls, we have $10^{10} \simeq 2^{33}$ i.e. $n = 33$ bits of information per soul. In the extreme case, suppose there are 10^{10} galaxies each with 10^{10} stars, each of which has a planet supporting life. If each planet sustains 10^{12} tons of biomass and all of it were bacteria of $1\,\mu$m dimensions and the same density as water (1 ton per m^3) there would be 10^{30} organisms per planet. If they all reproduced at 40-minute intervals (10^4 times a year) and have been doing so since the Big Bang some 10^{10} years ago, the total number of organisms to have existed would be $10^{(10+10+30+4+10)} = 10^{64} \simeq 2^{213}$. So if each had a separate soul it would need to carry 213 bits of information to ensure its individuality. I'll make it 220 bits — I'd hate to leave anybody out. This number is reduced if the same set of souls keeps going round, or if only higher organisms have souls; it is increased if each soul carries additional information (e.g. elements of a personality, memories of previous lives) into its new abode.

Can we measure this number? The informational interpretation of entropy associates one bit of information with energy $kT \ln 2$ joules, where k is Boltzmann's constant and T is the absolute temperature. So n bits implicate in their transmission at least $E = nkT \ln 2$ J. This is pretty tiny: with $T = 310$ K (body heat) and $n = 220$ bits it is only 6.5×10^{-19} J, weighing $m = 6.5 \times$ $10^{-19}/c^2 = 7 \times 10^{-36}$ kg or about 10^{-32} g, so there's no hope of weighing it. (There's an anecdotal claim — *World Medicine*, 11 September 1974, p. 107 — that the change of weight on death or conception is 1.8 g, but this seems clearly absurd.)

But presumably this energy must occur as a single photon, and its frequency will be $v = E/h = nkT \ln 2/h$ and its wavelength $\lambda = c/v = hc/nkT \ln 2$. And since the foetus gains order when it acquires its spiritual identity, its entropy is decreased, and a photon of this wavelength must be *emitted*. (Cf. the emission of latent heat when liquid water freezes to the more highly ordered ice.) Now this is much more hopeful. For $n = 220$ we find $\lambda = 300$ nm in the near UV; for $n = 33$, $\lambda = 2030$ nm in the near infrared. Clearly by detecting this photon of incarnation and measuring its wavelength we can determine the exact time of entry of the soul and its information-content. Fortunately human flesh is fairly translucent to light in the near visible region, so external photomultipliers stand a chance of intercepting it. But introducing, e.g., a quartz optic fibre into the womb gives a better chance if we can manage it.

Down with growth

DREADCO's Horticultural Department once trained runner-beans up semicircular poles to see if they would follow the curve back into the ground. This they declined to do, though they became noticeably confused on reaching the horizontal section. The experiment raises the question of how plants know which way is up. A current theory is that small grains (statoliths) in certain cells indicate the vertical by settling to the bottom. So Daedalus has been watering selected plants with heavy water, hoping to increase the density of the cell fluids sufficiently for the gravi-sensing grains to float to the top. In this way the baffled vegetables should dive back into the soil, with fascinating consequences. At first Daedalus saw the technique merely as a novel weed-killer, misleading dandelion and couch-grass seeds into sprouting suicidally downwards. But he soon saw the value of reversing plant growth at a later stage. Not only might this make possible the hand-harvesting of potatoes, onions, beetroots, mangel-wurzels and other root-crops, but new forms of plants could be made. Thus by intermittent application of high-density fluids, hops or runner-beans could be braided alternately up and down the same short length of pole, a great economy. Similarly, U- and S-shaped timbers could be formed for one-piece door frames and intricate joinery; and croquet-hoops could be grown *in situ*.

DREADCO's horticultural team is also experimenting with delicately balanced, subcritical applications of inverting fluids to enable crops to be grown at the proper angle on steep hillsides, or even horizontally from cliff faces. Indeed, Daedalus foresees a novel graviculture with vegetables planted on the floor, walls and ceiling of green-houses, and all growing towards a central light-source. And for households too sophisticated even for rubber-plants, the DREADCO ceiling-lawn should provide a really with-it conversation-piece!

(*New Scientist*, 28 January 1971)

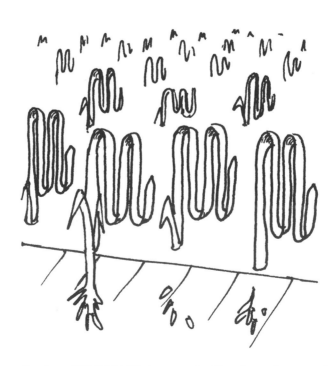

Intermittent application of DREADCO's inverting fertilizer keeps the grass short without inhibiting its growth. A lawn thus 'thickened' also limits predation by tripping up the birds

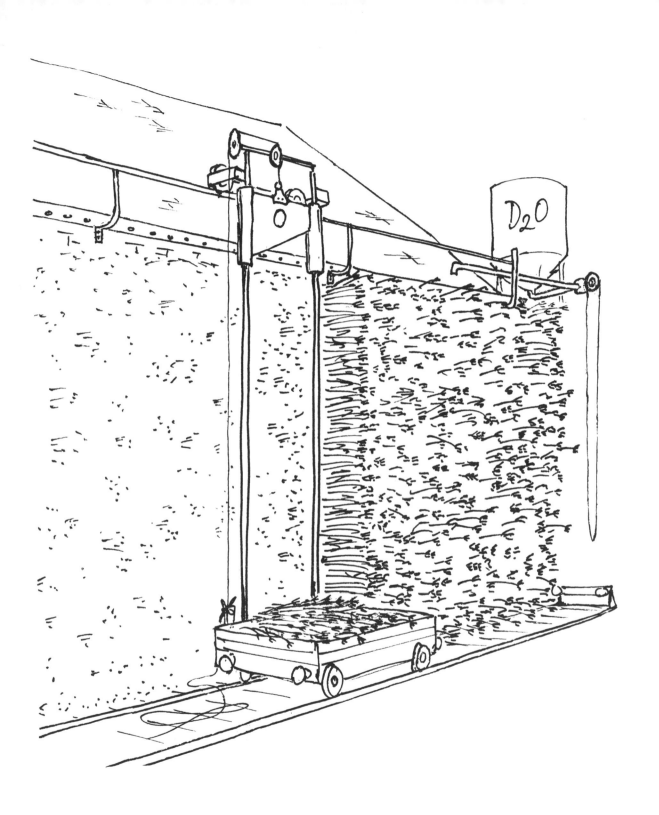

Arrangements for growing and harvesting horizontal wheat

Lasing the clouds away

The weather, in particular rain, is still annoyingly beyond human control. Even the most careful cloud-seeding cannot deliver its rain to any precise point. Daedalus plans to remedy this technological deficiency, and points out that if you ejected an electron from a cloud droplet, it would be left with a positive charge. The ejected electron would soon lodge in an adjacent droplet, which would acquire a corresponding negative charge. The two droplets would then rush together under their electrostatic attraction, and coalesce. Repeat the process within a cloud and all the droplets would grow steadily by repeated coalescence until they fell out of the cloud as rain. The obvious way of ejecting electrons from cloud-droplets is by using the photoelectric effect; Daedalus calculates that ultraviolet light of 100 nanometres wavelength or less is energetic enough to work this trick. Furthermore, there are so few droplets in a cloud compared to the photons in a UV beam that a mere watt or so of radiation should be ample for the job. So the DREADCO high-precision weather machine is simply a little steerable UV laser pointing at the sky.

At last the sombre British climate will be tamed. From the steady overcast, or the scattered clouds, the farmer will be able to cut out a swathe to fall precisely in his own fields. The ceremonial gathering or threatened garden fête could pre-emptively discharge just the required section from an approaching storm and steer it into a suitable canal or reservoir, leaving the celebration bone-dry while rain swept past its outer boundaries. A second laser aimed at the Sun could bore a hole through the murk of just the right diameter to bathe the proceedings in a precise spotlight of sunshine. The strange circles and oblongs thus carved from the cloudscape would look rather odd, but turbulence would soon heal them, so that communities downwind could employ the same tricks again. Indeed, this new form of sky-writing will enable advertisements, political slogans, and graffiti of all kinds to be transiently engraved into the scudding cumulus. But the sudden flash-flood rain-storms thus unleashed on the watchers beneath might dampen their enthusiasm for products or policies recommended in this striking manner.

(*New Scientist*, 4 December 1980)

From Daedalus's notebook

One photon of radiation should, on simple photoelectric theory, eject one electron from a droplet. It works in practice too — e.g. Millikan's classic determinations of e using radiation-charged droplets. The first ionization potential of water is 12.56 eV, i.e. $I_p = 12.56e = 2.0 \times 10^{-18}$ J; so electrons can just be ejected from water by radiation of frequency $v = I_p/h = 3 \times 10^{15}$ Hz, i.e. of wavelength $\lambda = c/v = 100$ nm, in the UV.

What intensity of such UV will be needed to condense cloud? Suppose a typical cloud is formed by the cooling of saturated air from 20 °C (when it can hold some 0.017 kg m^{-3} of water as vapour) to 10 °C (when it can only hold 0.009 kg m^{-3} as vapour). The surplus water condensed into cloud-droplets then amounts to $M = (0.017 - 0.009) = 0.008$ kg m^{-3}. If the droplets are (say) 3 μm across (i.e. $r = 1.5 \times 10^{-6}$ m) and have density $\rho = 1000$ kg m^{-3}, then each one weighs $m = 4\pi r^3 \rho/3$, and the number of drops per cubic metre of cloud will be:

$$
\begin{aligned}
n &= M/m \\
&= 3M/4\pi r^3 \rho \\
&= 3 \times 0.008/(4 \times \pi \times (1.5 \times 10^{-6})^3 \times 1000) \\
&\simeq 5 \times 10^{11} \text{ m}^{-3}
\end{aligned}
$$

Each UV photon hitting a droplet will eject an electron into another droplet, causing the two droplets to fuse into one. It will therefore reduce the number of droplets in the cloud by one. So n photons per cubic metre will fuse all the droplets and precipitate the cloud completely. So C cubic metres of cloud could be precipitated every second by a radiation-flux of Cn photons a second. If each photon has energy I_p, and (say) $C = 10^5$ m^3 s^{-1}, then the radiation power required for this feat is:

$$
\begin{aligned}
P &= CnI_p \\
&= 10^5 \times 5 \times 10^{11} \times 2 \times 10^{-18} \\
&= 0.1 \text{ W!}
\end{aligned}
$$

Even allowing for inefficiencies, a few watts only should precipitate vast amounts of cloud.

What rainfall will result? Suppose we aim our laser light upwards as a sheet-beam of width 100 m, and cloud 100 m thick sweeps over it at a velocity of 10 m s^{-1}. Then the beam is intercepting $C = 10^5$ m^3 s^{-1} of cloud as we assumed above, and precipitating it continuously as a swathe of rain over a front 100 m wide and (say) 1 m deep: i.e. the stimulated rainfall is concentrated on an area $A = 100 \times 1 = 100 \text{ m}^2$. Rain will hit this area at a volume-rate $\dot{V} = CM/\rho$, i.e. at a linear rate of

$$
\begin{aligned}
\dot{x} &= \dot{V}/A \\
&= CM/\rho A \\
&= 10^5 \times 0.008/(1000 \times 100) \\
&= 0.008 \text{ m s}^{-1} \\
&= 8 \text{ mm s}^{-1} \text{ or about 1100 inches an hour.}
\end{aligned}
$$

Pretty torrential! Clearly the technique could easily be adapted for filling reservoirs, canals, etc. With accurate servo-controlled aiming, it could also make firefighting a lot easier.

DREADCO's laser precipitator in action

Rusty armour

Musing on the neutron bomb — which kills people while leaving tanks and other weapons unharmed — Daedalus began to realize the advantages of a converse weapon: something that destroyed tanks while leaving people unharmed. He recalled that mechanical stress makes many materials remarkably vulnerable to corrosion. The corrosive molecules get into tiny cracks in the material, and react at the bottom of the crack where the material is most stressed and therefore most vulnerable, helping the stress to spread the crack. Now some molecules react with surprising changes of shape or volume. When an oxygen atom, for example, corrodes metal and becomes an oxide ion, its diameter nearly doubles. So DREADCO chemists are seeking vapours whose molecules expand or unwind very dramatically when they react. When they get into a microscopic crack in a structure and react at the crack bottom, the irresistible molecular expansion will drive the crack further by sheer wedge action. Thus these devilish stress-corrosives generate their own stress, and will shatter material on mere contact! Each of the DREADCO 'Shattergases' (Regd) will only touch the class of metal or plastic it has been chemically tailored to, and with luck most of them will be quite harmless to people.

This humane anti-weapon will revolutionize warfare. A shattergas tailored to armour-plate steel, reducing tanks and armoured cars to fragments around their baffled occupants' feet, and crumbling gun-barrels of all calibres, would be great fun. But more economical would be a selective shattergas attack, say on copper alloys (many of which are notoriously liable to stress-induced corrosion). Most military hardware would be left dumb and useless with all circuits opened and cartridge-cases split. And soldierly morale would collapse as brass buttons and insignia of rank crumbled, and trousers fell around combat boots amid splinters of belt-buckle. Battle would be reduced to an undignified punch-up between weaponless and half-dressed military clowns. Local civilians, instead of being terrorized, might even flock to watch it.

(*New Scientist*, 15 June 1978)

From Daedalus's notebook

As a crack spreads through a solid, it creates new surface (the sides of the crack); to propagate spontaneously, it must get from somewhere the energy needed to form this new surface. In a stressed solid, the energy has to come from the relaxation of the previously stressed material around the new crack. For small cracks and moderate degrees of stress, this energy-source is inadequate; so most engineering materials as usually employed are safe against brittle failure. But now suppose we consider an extra energy-source: the energy released in the corrosion of the new surface. A monolayer, of e.g. oxidation-product, must form on it almost instantaneously. Will this new source of energy tip the balance?

Most metals have a surface energy E_s of the order of $1\,\text{J m}^{-2}$; e.g. for iron $E_s = 1.7\,\text{J m}^{-2}$. The density of iron is $\rho = 7900\,\text{kg m}^{-3}$ and its molar mass is $A = 0.056\,\text{kg mol}^{-1}$. So iron contains ρ/A moles m^{-3} or $N = \rho L/A$ atoms m^{-3}. So an iron *surface* contains $N^{2/3}$ atoms m^{-2}, and hence a number of moles per square metre of:

$$
\begin{aligned}
M_s &= N^{2/3}/L \\
&= (\rho/A)^{2/3} L^{-1/3} \\
&= (7900/0.056)^{2/3} \times (6.022 \times 10^{23})^{-1/3} \\
&= 3.2 \times 10^{-5}\,\text{mol m}^{-2}
\end{aligned}
$$

The heat of rusting of iron to $Fe_2O_3 \cdot xH_2O$ is $-\Delta H = 2.7 \times 10^5\,\text{J mol}^{-1}$. So the energy released when an iron surface rusts to a monolayer of $Fe_2O_3 \cdot xH_2O$ is $H_s = -\Delta H M_s = 2.7 \times 10^5 \times 3.2 \times 10^{-5} = 8.6\,\text{J m}^{-2}$. So $H_s/E_s = 8.6/1.7 = 5$; i.e. the formation of a monolayer of rust on an iron surface releases about five times as much energy as was needed to form that surface in the first place. So if only 20% of this energy of rusting can be fed back into forming more surface by crack-propagation, the crack will run spontaneously even in unstressed material. And if the rapid corrosion of new surface extends to several monolayers — as it almost certainly does — then the energy supply is even more copious and still less need be fed back to make a crack spread.

The density of $Fe_2O_3 \cdot xH_2O$ is about $3000\,\text{kg m}^{-3}$, whereas the density of iron itself is $7900\,\text{kg m}^{-3}$; i.e. there's a 2.6-fold expansion when iron rusts. If instead of water-vapour, we put into the air the vapour of an even bulkier 'molecule of crystallization', we could increase this expansion considerably. The corrosion-layers on either side of the crack would then expand in their formation until they touched. Further corrosion would then exert wedge action on the crack to spread it. If corrosion forms n monolayers on the surface, then this mechanism need only feed back $(20/n)\%$ of the energy of corrosion to make the crack run away. The corrosion-expanding vapour would then constitute a 'shattergas' for iron or any other metal of similar corrosion chemistry.

Crack propagation by expansive corrosion

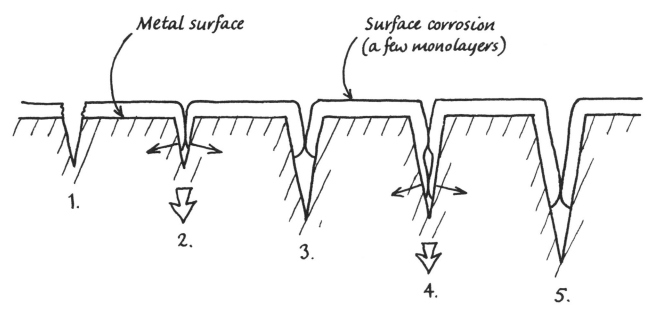

Metal surface

Surface corrosion
(a few monolayers)

1.
2.
3.
4.
5.

1. New crack forms
2. Corrosion in crack wedges sides apart
3. Crack deepens till corrosion is accommodated
4. Corrosion commences on the newly-made surfaces of the extended crack, and wedges them apart
5. So crack deepens again...

(In practice the process should be continuous rather than step-wise)

Daedalus comments

This proposal was initiated by a letter I received from a Frau McClanahan in West Germany. She wished that the likely disruptive impact of World War III on local civilian life could be blunted in some way. Her suggestion was to employ infrasonics to destroy metal weapons; she wrote:

Dear Daedalus,

Do please try to come up with an alternative to the Neutron Bomb before someone settles the Geschrei by reducing me (among others) to a fine stain on my kitchen floor. Perhaps a start could be made on an 'Anti-Bomb' which reduces metals to powder but leaves people untouched. The sonic research some time past of the Frenchman Gavraud could be continued in this direction, perhaps by tuning in on the molecular oscillations of, say, metals.

We wouldn't have to use giant Gendarme peawhistles for the delivery, but that might provide a bit of elan and morale for our side. The correct oscillation to disintegrate metal, moreover, might leave the hapless soldier minus belt-buckle as well as tank; the resulting need to hold up his trousers could prevent further mischief, say at the stone-throwing level . . .

My feeling was that infrasonics wouldn't work — the wavelengths are all wrong, and precise tuning would be very difficult anyway. But Daedalian gallantry could not ignore this touching plea from a damsel in distress, and I tried to develop a weapon that kept as close as possible to the spirit of her brief.

A new perspective on comets

Daedalus has a theory of comets, those enigmatic entities which so puzzle astronomers. They have long, highly eccentric orbits (indeed, many come near to escaping from the Solar System altogether); they lose a lot of material boiled off as a tail every time they traverse the close solar-approach section of their orbits; often they appear to vanish entirely after this stressful solar encounter; yet the supply of comets never diminishes. Where do they all come from?

Daedalus points out that the interstellar gas, that tenuous medium in which the stars and planets swim, is by its vast volume the predominant constituent of the Universe. And a rocky asteroid with a highly eccentric orbit would spend by far the greater part of its time in the freezing cold of outer space, far from the Sun, ploughing through this material. So, says Daedalus, such an asteroid would act as a condensation-nucleus and build up a thick coating of condensed interstellar matter — water, ammonia, methane, even hydrogen. When after decades in the outer darkness it returned to swoop briefly round the Sun again, this volatile material would boil off in the sudden heat, and be blown away in the solar wind as the familiar cometary tail. Furthermore, this slow accumulation and rapid loss of matter must strongly perturb its orbit. An astronomer predicting the return of such an entity on the familiar orbital assumption of constant mass would get it wrong. When it didn't turn up on time he would assume it had vanished, and would hail it as a new comet when it subsequently appeared on an unpredicted course. So the whole problem of cometary supply is an illusion. New comets are simply old 'vanished' comets coming back with new overcoats and along unexpected orbits. Indeed, thinks Daedalus, there may be only *one* comet. Sometimes its orbital whims drive it deep into gas-rich space so that it returns in evaporative splendour; other times it picks up little gas and is hardly visible on its return. Meanwhile, astronomers continue to catalogue it under ever more fictitious names . . .

This theory has exciting implications. If comets really are interstellar dustpans sweeping up and bringing back material from far beyond the planets, their spectroscopic analysis should brilliantly reveal the contents of deep space. Their observed large size and complex composition already show that the interstellar gas must be rather denser than is commonly assumed, and must contain a lot of interesting molecules — as microwave astronomy confirms. So Daedalus would like to clinch matters by getting his hands on some of it; and is designing for this purpose a space-probe based on his cometary theory. Any space-probe protected from the Sun's rays must cool towards the ultimate cold of space, only a few degrees above absolute zero. So Daedalus's probe has a big metallized reflective polymer-film 'sunshade' many kilometres across. In its shadow a huge polythene-film cylinder, inflated with very low-pressure hydrogen or helium, acts as a cryogenic surface on which the molecules of space condense. The weightlessness of space makes such vast, frail, feebly inflated structures entirely feasible. The interstellar condensate on the cylinder will make an impalpably thin film, even though the huge area of condensing surface will collect it at quite an impressive total rate. So the cylinder will be rotated slowly, passing its surface over a radioisotope-warmed roller to distil the accumulating condensate continuously onto an endlessly moving belt. This will carry the material to two collector-terminals, which will distil it from the belt surface as a concentrated stream. If the spectroscopic analysis of comets is anything to go by, the product should comprise mainly water, ammonia, methane and other small molecules. These are useful materials to accumulate from space; rocket fuels can be made from them, and Daedalus's space-probe will sacrifice some small proportion of its 'catch' as venting-gas for the stabilization-thrusters. In the very long term space-condensate might even free the chemical industry from its dependence on oil. But as a first experiment, Daedalus wants to use space-condensate to test his cometary theory. He aims to assemble it round a rocky core as an artificial comet, and then fire it into an appropriate orbit to compare it with the real thing.

(*New Scientist*, 2 and 9 September 1976)

Daedalus comments

I am pleased to report that I am not alone in recognizing the disturbing influence of evaporation on cometary orbits. Paul Weissman of the Jet Propulsion Laboratory (*Astronomical Journal*, Vol. 84, 1979, p. 580, as reported in *New Scientist*, 28 June 1979, p. 1091) has now developed a very subtle theory involving the direction of spin of the comet's nucleus. If it is spinning one way, evaporation during its solar passage elongates its orbit; if it is spinning the other way, the orbit is contracted. But so far, nobody seems to have developed the complementary theory of mass gain over the outer loop of the orbit.

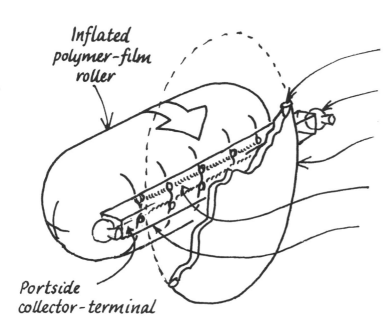

Inflated polymer-film roller

'Inner-tube' inflated to open sunshade

Starboard accumulator and thruster

Sunshade (two metallized films in vacuum)

Belt system to strip condensate from inflated roller

Box-girder frame (width vastly exaggerated)

Portside collector-terminal

Detail of portside collector-terminal

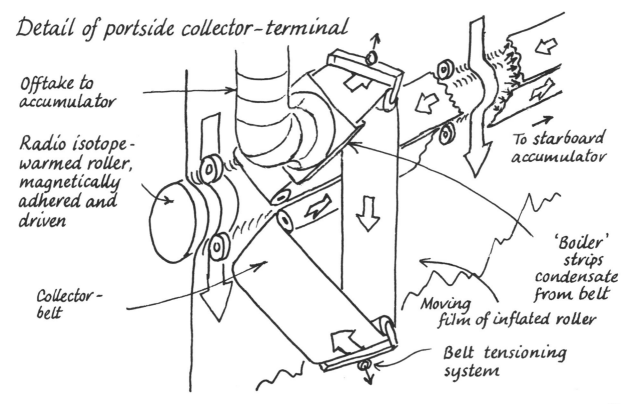

Offtake to accumulator

Radio isotope-warmed roller, magnetically adhered and driven

Collector-belt

To starboard accumulator

'Boiler' strips condensate from belt

Moving film of inflated roller

Belt tensioning system

Sloping water

How much more useful canals would be, if they were not obstructed by locks at every change of level! Daedalus points out that a dense liquid in one limb of a U-tube will balance a taller column of a less dense one in the other limb. Similarly, if a number of tubes are conjoined, and progressively denser liquids are placed in successive tubes, a steady decrease in level along the series will result. So in the limit, he argues, if a liquid of steadily increasing density could be stabilized, it would exhibit a sloping surface. Daedalus's first notion of merely heating the water at one end to lower its density by thermal expansion would give only a small effect, and would anyway be complicated by convection. He therefore proposes to stabilize a sloping surface by suspending dense magnetic particles in the water, and attracting them preferentially to one end of a lock by powerful permanent magnets. Variations of level are indeed observed in such liquids when placed in a magnetic field. The new locks would allow uninterrupted passage of shipping both up and down hill, though vessels would experience a sudden change of apparent buoyancy on entering the magnetic liquid, which would have to be divided from the continuation of the canal by a flexible underwater partition.

Many other uses suggest themselves for Daedalus's sloping-water system. In particular, it could enhance the capacity of reservoirs: if surrounded by magnetic densifiers, the water could be heaped up in the middle much above the shore-line. Such treatment of existing reservoirs could multiply our existing water-storage manyfold, and make expensive new reservoirs unnecessary. Novel sports like gravity water-skiing would also become possible. But what the ducks would make of it remains to be seen.

(*New Scientist*, 1 September 1966)

From Daedalus's notebook

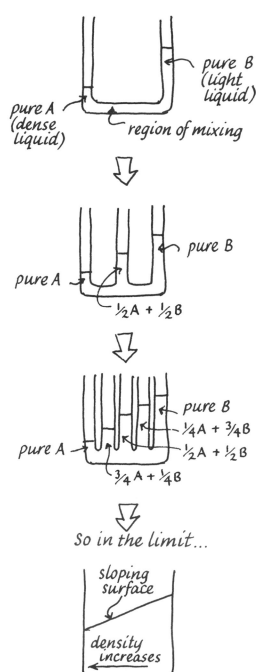

pure B (light liquid)

pure A (dense liquid)

region of mixing

pure B

pure A

½A + ½B

pure B

¼A + ¾B

½A + ½B

pure A

¾A + ¼B

So in the limit...

sloping surface

density increases

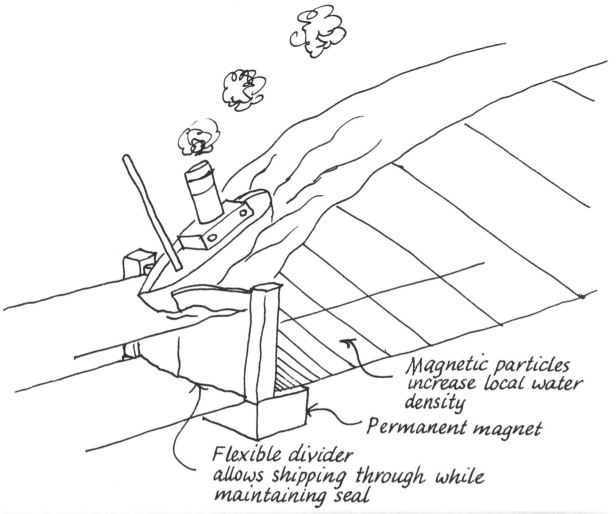

Magnetic particles increase local water density

Permanent magnet

Flexible divider allows shipping through while maintaining seal

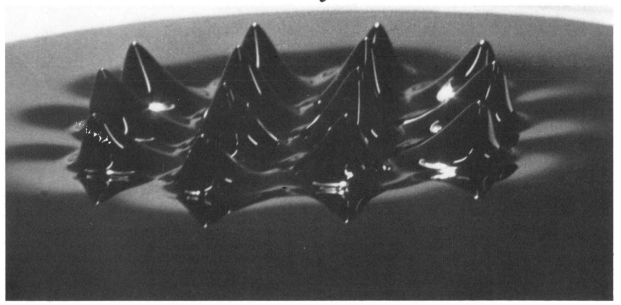

Level variations in a magnetic liquid in the presence of a complex vertical magnetic field. (Courtesy J. Popplewell and S. W. Charles)

The fields of sleep

Daedalus is a 'late bird'. Indeed he is barely conscious before about 10 a.m. although he has trained his internal autopilot — presumably some part of the spinal cord — to give a sluggish semblance of responsiveness before then. His envy of early risers who can even do useful work before about midday matches his pity for the millions of people hopelessly ill-fitted to the organizational 9-to-5 straitjacket. Daedalus wonders how these persistent internal time-patterns keep in step with official clock time. He suspects that his own rhythm is slaved to the radio news bulletins, and is experimenting with an alarm-clock-cum-tape-recorder to tape them silently and play them back at the wrong hours to see how he reacts. But daylight and BBC transmissions are not the only phenomena which can control the phase of a sleep/wake cycle. Temperature-alternations, a.c. electromagnetic fields, presence or absence of specific background sounds, all can act as time-cues, 'zeitgebers' as they are called, to govern circadian rhythms. Daedalus recalls recent experiments in which the day/night rhythms of mice, normally governed by light intensity, were induced to follow an electric field which was switched on every 12 hours and off for the next 12 hours. Under constant illumination, the creatures took the presence or absence of the field as their criterion for day or night. Daedalus is hoping that the same phenomenon in man will prove an ideal way of rephasing human rhythms. His 'electro-bed' has sensors to detect when it is occupied, and whether its occupant is awake or asleep; it switches the field on during the sleeping period so that the occupant subconsciously comes to associate it with sleep. Meanwhile a small computer logs the accumulating sleeping record and establishes the pattern of the user's rhythm. When this is accurately known, the electro-bed begins to edge the field-on period towards the desired pattern (say 11 p.m. – 7 a.m.), and the sleeper finds his habits altering towards this socially useful rhythm. Thus the enduring ordeal of misphased organization men could be ended, and nocturnal harmony brought for the first time to ill-assorted married couples. The electro-bed could also be programmed to adjust shift-workers to their changed circumstances, and (as a portable 'electropillow') to enable airline crews and flight-dazed executives to adapt better to their bewildering lives.

(*New Scientist*, 15 March 1973)

Daedalus comments

The experiments with mice to which I referred in this item were performed by H.B. Dowse and J.D. Palmer of the Biology Department of New York University, and were described in *Nature* in 1969 (Vol. 222, 10 May, p. 564). In a rather naive and biological manner, these workers exposed their mice to the entraining electric field by raising the complete cage to 500 volts on a 24-hour cycle. The mice, being *inside* the cage, will have experienced very little of this; only small and bafflingly inhomogeneous fields will have existed around the edges of the cage and between the wires. The main effect will have been felt *outside* the cage, in the form of lethal danger to unwary mouse-lovers seeking to pet the creatures or push tasty morsels for them through the cage wires. Nonetheless, when the field was applied simultaneously with light on a 24-hour cycle, most of the mice became entrained by it. When the lighting cycle was discontinued and they were pro-vided with constant dim illumination, leaving the field as their sole day/night indicator, they remained in good synchronism with it. Their alternations of activity and sleep kept phase with the field throughout the duration of the experiment. Interestingly enough, not all the mice were entrained; and some that were entrained adopted wake/sleep rhythms with a distinct phase-shift with respect to their guiding field. The electro-bed will have to incorporate provision for recognizing and allowing for this complication.

That's entertainment, mathematically speaking

Daedalus is quite alarmed by the steady pressure for more and more TV channels. It is not as if there were any vital demand for this extra variety — most viewers actually use TV as a completely unmeaning diversion, like looking out of a train window or watching the clothes in a tumble drier. And the expense of providing new channels, with their extra transmitters and relay-links, and the wretched plight of the luckless hacks who will have to invent ways of filling them, prompts him to suggest an easier way out. He recalls a computer-analysis of the Homeric sagas which neglected all considerations of plot and long-range structure, and assumed that each incident in a theme was selected by the poet spontaneously, by a weighted random choice from the options open to him at that moment. Many TV programmes already give the impression of being put together in this hand-to-mouth manner, and the average inattentive viewer will hardly notice overall plot-development anyway.

So Daedalus is computer-analysing a large amount of TV purely statistically, to tabulate the chances of any one scene being followed by any other. All these probabilities, together with a large selection of TV scenes, will be stored on a computer disc-file. Daedalus's 'Stochastic Entertainment Generator' (SEG) will then generate endless typical TV material just by selecting a scene at random from its stock of 'programme beginnings', using the stored probabilities to make a weighted random choice of its successor, presenting that scene, making a further weighted random choice of *its* successor, and so on: until it finds it has selected a recognized 'terminator' scene (e.g. a happy ending). Thus from moment to moment the programme will proceed much like typical soothing TV, with nothing to alarm or worry the viewer: surprising developments or juxtapositions (in the shape of low-probability choices) will be rare. Some internal consistency will be needed, however. SEG will use advanced pattern-manipulation techniques to ensure that the moving elements in each scene (people, cars, etc.) are selected from a specific 'cast' which, if not constant, at any rate changes only slowly over time. And minimal long-term planning will be necessary. Thus if SEG introduces an 'advertisement' scene, it must remember to provide an ad-terminator and recapitulation scene shortly — otherwise the next random choice might introduce a second advertisement within the first, even stacking them many levels deep, or just developing the rest of the programme from the ad. SEGs with a few thousand basic scenes (each with

as many micro-variations as possible) should simulate a TV channel adequately enough to be sold as plug-in TV accessories. They should effortlessly satisfy the mass of undemanding viewers, leaving the existing channels open to cater for that small fraction of people who actually want to watch something specific in earnest.

(*New Scientist*, 20 May 1976)

Daedalus comments

The computer-analysis of the Homeric sagas referred to in this piece occurs in *Thematic Structure in Homer's Odyssey* by P. V. Jones (PhD thesis, King's College London, 1971, pp. 298 *et seq.*). It considers Homeric composition as a first-order stochastic process, i.e. it assumes that Homer has no memory of the previous development of his story, but chooses each successive element purely on the basis of the possibilities which open up from the one he is currently reciting. This isn't a bad representation of the predicament of an oral reciter, and lends itself to computer simulation; in this case, the results seemed to show that this Homeric model was essentially correct. (In fact, Homer does remember enough of the story so far to avoid wandering in circles and repeating himself; the computer simulation occasionally did so.)

Programme material such as TV could certainly be analysed and simulated in this way with little difficulty in principle. But its actual implementation on computer and video hardware is much more feasible now than when I put the idea forward in 1976. The advent of frame-stores capable of holding a complete TV frame in digital form, and processing it in real time, makes possible the continuous assembly of TV images from separately stored 'cast' and 'background' images, as well as ringing the changes on picture-enlargement and composition, left/right reversal, and so on. In this way a modern SEG could assemble from its basic finite library so many different images and scenes that the limits on its repertoire would take some time to become apparent even to an attentive viewer. But the incredible popularity of the new TV games suggests that even a rudimentary visual repertoire can be endlessly entrancing provided the viewer has some control of it. So by allowing the viewer access to the stored probabilities, permitting bizarre forced choices of plot development, an interactive SEG could provide unlimited gripping entertainment from even quite a small basis set of scenes and characters.

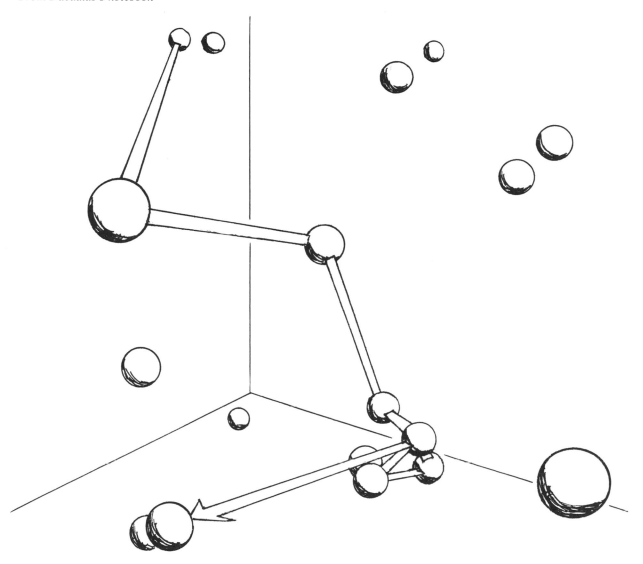

All possible TV scenes can be located in a multidimensional 'image space'. With an appropriate choice of dimensionality for this space, the scenes can be so positioned that the probability of scene (a) being succeeded by scene (b) is some inverse function of their distance apart in the space. Very common types of TV scenes (eg. ones dealing with sport, violent crime, etc.) will occur as dense clusters in image space; rare and uncharacteristic scenes will occur as isolated points. Any possible TV programme can be represented by a 'walk' in this image space. The SEG-machine will generate programmes by performing a 'random walk' in this space, weighted so as to make short jumps between close neighbours more likely than longer jumps.

Mind and antibody

Daedalus sympathizes with those feminists who emphasize the fear of rape in limiting the social freedom and equality of women. DREADCO's existing anti-rape device — imitation male genitals in realistic plastic — undoubtedly gives the rapist serious second thoughts, but is rather a nuisance and has to be worn in anticipation. So Daedalus is working on a new defence. He points out the extraordinary efficiency with which an alien substance in the body (an antigen) is mopped up by the appropriate defence chemical (antibody). DREADCO biochemists are seeking animal species — guinea pigs, parrots, armadillos or whatever — to which human male sex hormones are alien substances. When injected with such hormones, the creatures will raise defensive antibodies against them, and these can later be extracted from their bloodstreams. Now the fastest way to get a chemical into the body is by inhaling it as an aerosol (which is why smoking and glue-sniffing are effective). So DREADCO's new anti-rape device is an aerosol can containing these animal antibodies against the male sex hormones. Sprayed into the air, the vapour-cloud will inevitably be inhaled by all nearby, and will almost instantly remove all the male sex hormones from their bloodstreams. For women this will not make much difference, but the whole biochemical underpinning of masculine sexual arousal will be removed. The rapist's erection will collapse and his interest in the whole business will flag until, in due course, his body has synthesized enough new hormones to use up the inhaled antibodies and restore normal levels.

Some psychologists hold that sex is the ultimate drive behind all masculine aggression and enterprise. If so, the DREADCO desire-collapsing aerosol will reduce the fiercest man to a sort of gormless neutered-tabby complacency. It would then be the ideal non-violent countermeasure against terrorists, hijackers, and violent criminals generally, who tend to be predominantly male. Daedalus did wonder whether he had also placed a terrible new weapon in the hands of the extremists of the feminist movement. But since both sexes have a minority component of the other's sex hormones in their system, the aerosol will presumably also subvert extremes of aggression in women too.

The principle of this neat and humane device can clearly be extended. It can in theory remove from the body any substance complex enough to appear alien to the biochemical defence systems of armadillos, or some other suitable animal species. This species could then be used to raise antibodies to the substance in question, to be extracted for aerosol use. The really attractive feature of the scheme is the speed with which it acts: inhaled substances get round the body in seconds. For the main trouble with tranquillizers, pep-pills, sedatives and hallucinogens is the time they take to wear off. An aerosol containing antibodies to such compounds, capable of clearing them from the system almost instantly, would be a tremendous boon. To be able to escape from the humdrum world into a hallucinated dream, and yet after an hour to be able to snap back immediately to rational alertness, would reduce LSD to something as manageable as ITV, and vastly more entertaining. So Daedalus is devising derivatives of LSD, Mogadon, etc., bearing sidechains of the sort to which antibodies can be raised. By injecting these derivatives into the hapless armadillos, etc., he will obtain the appropriate antibodies needed to neutralize them. He will then sell these new psychoactive drugs along with a matching aerosol alarm-clock. You just set the dial before popping your pill. After a predetermined period of expanded consciousness, or merely deep sleep, the machine automatically delivers its healing spray; and at once you are returned to the clean daylit world, alert, refreshed, and ready to go.

The major application of the technology should be to travel. Higher and higher speeds, with their huge capital and energy costs, are quite irrelevant to so-called business efficiency: they are really craved just to keep businessmen from being bored. Give the long-distance traveller by rail or air a sleeping-tablet or hallucinogen, and an aerosol-clock to wake him cleanly at his destination, and the whole drive for high velocity becomes unnecessary. Airlines, railways and bus companies would much prefer unconscious or flaccidly freaked-out passengers to the fractious mob they handle at present. The only loser would be the present, primitive drug technology thus outdated: the railway bar or airline drinks trolley.

(*New Scientist*, 27 October and 3 November 1977)

Daedalus comments

Since this was written, the introduction of the monoclonal-antibody process, which uses hybrid cells in culture to make pure antibodies in bulk, has made the scheme much more feasible and more humane. I was never happy about giving those poor armadillos hallucinations.

DREADCO's 'Feline Passion-killer' contains antibodies to the cat sex-hormones

Weight-whirlers

The human body is a marvel of feedback control. It manages to maintain constant temperature, and so on, no matter how its surroundings vary. Daedalus recalls that a high temperature is treated, not by forcibly cooling the patient, but by keeping him extra warm — just as one can correct a sticking thermostat by heating it still further till it ultimately clicks over to 'cooling'. This principle can clearly be extended, particularly to help those unfortunates who despite any amount of dieting remain depressingly overweight. Daedalus sees this as a malfunction of some internal 'weight-stat', and wonders how the body knows how heavy it is. At first he advocated an anatomical search for pressure-sensors in the heels and buttocks, signalling standing and sitting loads, but now recognizes the likelihood of some more general load-sensitivity. Any method of increasing the apparent loading should switch the weight-stat to 'reduce', causing effortless slimming.

The obvious method is to use centrifugal force. The technology of man-carrying centrifuges and rotating restaurants is already far advanced, making simple the design of a rotating health-clinic whose centrifugal force would expose the patients to a steady load of (say) $1\frac{1}{2}$ gravities. Such an extra apparent 'weight' should certainly do the trick. This implies a mere four revolutions per minute for a 50-metre radius clinic built on a railway turntable; individual rooms could be pivoted to maintain a constant apparent 'vertical'. And for slimmers who wish to modify their metabolism without the bother of attending a clinic, Daedalus is inventing a rotating dining-room chair to reduce the appetite at the crucial moments.

One unexpected corollary to this reasoning is the converse hazard to long-term space-travellers. In free fall their weight-stats would be so firmly stuck on 'increase' that they might weigh 30 stone on arrival at Mars.

(*New Scientist*, 15 February 1968)

From Daedalus's notebook

Medical opinion seems to be that anyone more than 10% over their 'design weight' needs treatment. So we need to convince the body that it weighs at least 10% more than it really does, i.e. the centrifugal clinic must produce 1.1g minimum. It's hard to stand and move about at 2g, so 1.5g seems a reasonable maximum. A simple triangle-of-forces calculation shows that this implies horizontal acceleration of 1.12g, and gives the

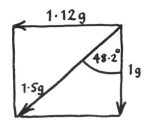

resultant 1.5g at 48.2° to the vertical. Hence the centrifugal acceleration $a = r\omega^2 = 1.12g = 11\,\mathrm{m\,s}^{-2}$. So a clinic with $r = 50$ m needs to spin at $\omega = \sqrt{11/50} = 0.47\,\mathrm{rad\,s}^{-1}$ or 4.5 r.p.m. Quite leisurely really!

The best design for the clinic would be a long corridor in the shape of a parabola, chosen such that wherever the patient stands he experiences the local 'gravity' at right angles to the floor, but the further out he goes the higher this 'gravity' becomes. At the full 50-m radius he'll be at 48.2° to the vertical. Then we can move patients about between the regions of higher and lower 'gravity' and discover what the medical effects are. The entrance should be inside the central annular thrust-bearing, so that staff can come and go without arresting the spin (cf. the ammunition-hoist in a swivelling gun-turret). The whole thing will be expensive, but since hundreds of millions are already spent annually on slimming with essentially no effect whatever, an effective treatment ought to pay its way handsomely!

Daedalus comments

The ideas put forward in this column were rapidly overtaken by events. Barely a month after it appeared, the American magazine *Time* (8 March 1968, p. 52) carried an account of some NASA experiments showing weight-losses in centrifuged rats; and long-orbiting astronauts have indeed shown marked gains in body-weight.

I later discovered that echoes of my reasoning had previously turned up even in mainstream physiology. Dr R. Passmore (*Penguin Science Survey B*, 1964, p. 144) gives an account of some experiments carried out by Sir Charles Dodds, who later became President of the Royal College of Physicians. Sir Charles tried weighing down experimental animals and human subjects with lead, and lightening them with balloons, to discover how this might affect their food intake and body weight. He reported that the only result of his experiments was that his colleagues began to doubt his sanity! Fortunately I no longer have to worry about such considerations . . .

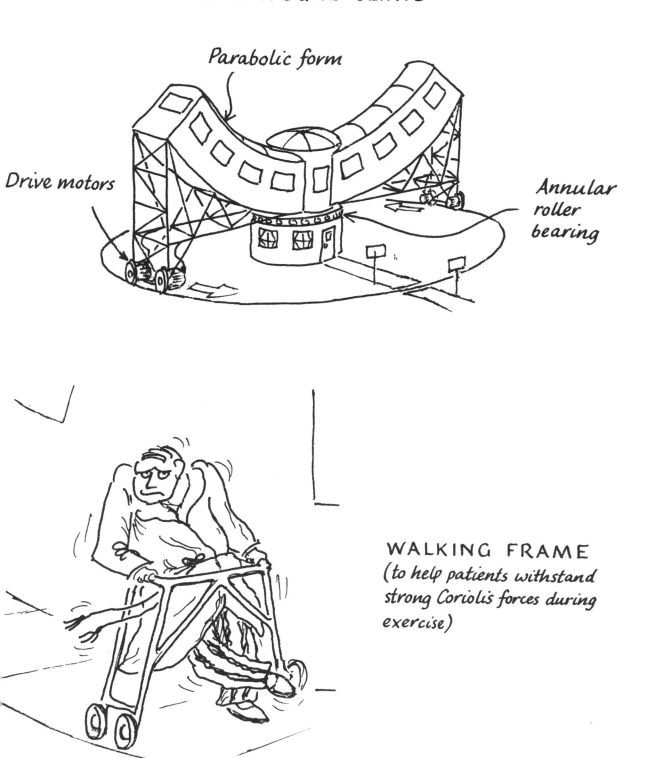

CENTRIFUGAL CLINIC

Parabolic form

Drive motors

Annular roller bearing

WALKING FRAME
(to help patients withstand strong Coriolis forces during exercise)

The uhuh-machine

A subtle facial code accompanies all normal conversations. It conveys such messages as 'I'm stopping now' or 'Do you want to comment?' Conversations over the telephone replace this code by an audible one, which uses little noises — 'uhuh', 'mmmm' and the like. Their importance becomes obvious if you stop making them. It usually takes only a few seconds for the other party to become anxious, and soon he will say 'Are you still there?' Only a very egotistical or strong-willed character will talk into a verbal vacuum for as long as a minute without becoming suspicious. Daedalus is often the victim of incredibly boring telephonic monologues, and has set DREADCO psychologists the task of cracking the 'uhuh' code, so as to imitate it on a computer. Verbal signals, like the rise in voice-tone at the end of a question, will be spotted by electronic filters and used to trigger appropriate recordings of 'yes, yes', 'really?', and so on, to keep the talker happy. The unwilling listener will be able to walk away and leave him at it. Conversational demands beyond the computer's small-talk will either request human aid, or will produce recordings of bacon-frying noises, metallic tapping and other Post Office specialities to confuse the issue. Daedalus foresees a brisk sale for this device, whose only drawback would be the risk of an endless, inane, expensive conversation if two of them ever got talking to each other.

The uhuh-machine also has creative potential. Used to generate encouraging sounds into headphones, it could prevent that drying-up which afflicts broadcasters and tape-fans confronted by an unresponsive microphone. Daedalus in fact gets many of his most brilliant ideas while talking to friends who understand nothing of the technicalities, but by acting as a sympathetic audience succeed in drawing out notions which might otherwise have remained latent. So several uhuh-machines will be installed in DREADCO's laboratories, for people who want to talk round problems, expound brilliant but unsound ideas without giving them away, or even grumble about the management. The ever-sympathetic machines should boost morale considerably!

(*New Scientist,* 14 March 1974)

Daedalus comments

An audience can be very stimulating. Eric Laithwaite remarked, in an interview with Anthony Curtis:

Of the inventions I've patented, I've discovered all but one when I've been talking to someone as I am to you now. If I meet a person I know is really interested I feel as if I suck knowledge out of him — he doesn't know it but I do. I have to put things more clearly for him than I would have for myself, and in that clarity I say something different.

(*New Scientist*, 20 September 1973)

Laithwaite calls this sympathetic idea-stimulation 'talking into a matching impedance'.

It doesn't take much response to produce this effect quite strongly. Since I described the uhuh-machine, the fascination of playing with computers has begun to spread relentlessly through the community. The reason is clear. A computer seems to have a personality — a narrow, obsessive, self-righteous personality, but a strong one nevertheless. And when the program is designed, however simplistically, to mimic human responses, the effect can be irresistible. When Joe Weizenbaum of MIT devised his program 'ELIZA', which bluffs its way through the role of psychiatric therapist, he was horrified to discover that many clients took the thing seriously, and demanded to be alone with the terminal to pour out their troubles! They wouldn't believe the machine was merely juggling words and had no idea what was going on.

So an interactive voice-computer only a bit more advanced than the uhuh-machine might be enormously popular, not only with inventors in need of creative stimulation, but also with ordinary people seeking the sort of sympathetic, uncritical acceptance and attention they now seek from cats or dogs. Combined with a refined version of the Facoder (p. 8) to provide an appropriately alert and sympathetic human face, it might prove the ideal 'technical fix' for loneliness, neurosis, alienation and boredom the whole world over.

The desert waterer

Daedalus has a novel scheme for extracting water-vapour from the desert air, based on the fact that sulphuric acid and golden syrup both take up water if you leave the top off the jar. In fact any solution whose vapour-pressure is less than that of the water-vapour in the air will take up water-vapour and dilute itself. But how to get this water out again? The obvious technique is to squeeze it out by reverse osmosis. So Daedalus began to invent a sort of sulphuric-acid-based water-press for desert travellers, in which water was squeezed out of the acid through a semi-permeable membrane, subsequently letting the acid absorb more water from the air when the pressure was relaxed. But he then reflected that hydrostatic pressure would do the trick just as well. A tall column of sulphuric acid would extract water continuously from the air at the top; this would diffuse or convect downwards and be continuously extruded under the great hydrostatic pressure from a semi-permeable membrane at the bottom. Daedalus was a bit worried about all that acid, and in any case golden syrup, with its higher molecular weight, would be better. It would need a sulphuric acid column some 2.4 km high to extract water from desert air of 20% relative humidity; whereas with golden syrup a mere 720 metres would do the trick. The water-miscible liquid with the highest molecular weight of all is probably 'Carbowax®'*, for which a trivial 50-m column would suffice. The energy for the separation ultimately comes from the fall of the water molecules so the process is continuous and automatic. So Daedalus's 'Desert Waterer' is a tall column of golden syrup or Carbowax, conveniently attached to the oil rig or transmitter mast or whatever brings you to the desert to begin with. A portable version supported by guy ropes or a balloon would serve the weary Bedouin on his travels, and the syrup (and just possibly the Carbowax) would be a useful emergency food supply. But attempts to reclaim the whole desert by vast forests of desert-waterers might be rather expensive.

(*New Scientist*, 25 May 1978)

*Carbowax® is a Union Carbide trademark for their liquid poly-ethylene-oxide.

From Daedalus's notebook

Suppose our solution contains N moles of water, molar mass M, mass therefore NM; and n moles of solute, molar mass m, mass therefore nm. Its total mass $W = NM + nm$; if the solution has density ρ, then its volume $V = W/\rho$.

The osmotic pressure by which this solution would seek to imbibe pure water through a semipermeable membrane is:

$$\Pi = nRT/V$$
$$= nRT\rho/W$$

Now if this solution is in a column of height h, the hydrostatic pressure at the bottom is ρgh. If pure water is to be expelled through a semipermeable membrane at the bottom, this pressure must equal or exceed the osmotic pressure:

$$\rho gh = nRT\rho/W$$
$$h = nRT/gW$$
$$= \frac{RT}{g}\frac{n}{(NM + nm)}$$

We shall need the top of this column to absorb water from a desert atmosphere of maybe only 20% humidity. So it must be a strong solution; Raoult's Law requires that its vapour-pressure should be less than p in

$$p/p_0 = 20/100 = N/(N + n)$$

This is true if $n = 4N$, making the height of the column

$$h = \frac{RT}{g}\frac{4N}{(NM + Nm)}$$
$$= 4RT/(M + 4m)g$$

Let's take $T = 300$ K and find h for three possible solutes. The molecular weight of water is 18, i.e. $M = 0.018 \text{ kg mol}^{-1}$. Sulphuric acid has m.w. = 98, i.e. $m = 0.098 \text{ kg mol}^{-1}$; sucrose syrup has m.w. = 342, i.e. $m = 0.342 \text{ kg mol}^{-1}$; Carbowax polyol has m.w. = 5000, i.e. $m = 5 \text{ kg mol}^{-1}$. We get, respectively:

2430 m (sulphuric acid)
720 m (syrup)
50 m (Carbowax)

Clearly Carbowax is the one to go for.

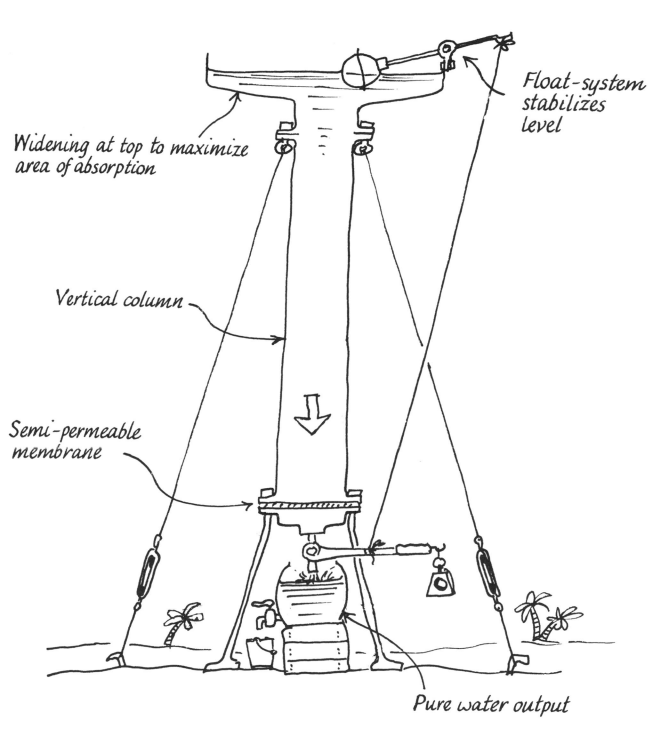

Widening at top to maximize area of absorption

Float-system stabilizes level

Vertical column

Semi-permeable membrane

Pure water output

THE DESERT WATERER

The eco-dirigible

The decline of social cohesion is well marked by the widespread fantasy of retiring from the rat-race and adopting some self-sufficient life-style. This fantasy is titillated by many little publications purporting to reveal how to make e.g. a windmill, a solar-power plant or a methane generator — though of course nobody ever actually builds anything to these designs. In keeping with this fantasy, Daedalus now presents the ultimate retreatist drop-out dream: the eco-dirigible. This depends on the simple fact that methane is lighter than air. The fermentation of cellulosic vegetable matter produces methane and carbon dioxide which are readily separable. Accordingly, a light enough fermenter could lift itself by filling a big gas-bag with methane. The airborne fermenter would be supplied with cellulose on the window-box principle, by putting the whole dirigible envelope 'under cultivation'. At first Daedalus dreamt of a vast gas-bag-cum-flannel sprouting mustard-and-cress and regularly mowed by magnetically adhering lawnmowers, but more sanely he has now devised a double-skinned transparent balloon-envelope through which a *Chlorella* culture is pumped. *Chlorella* is one of those algae whose photosynthetic uptake far exceeds that of conventional crops; its absorption of sunlight provides the primary energy-source for the dirigible. It feeds the fermenters, whose methane fills the gas-bag for lift and whose carbon dioxide is recycled for photosynthetic uptake by the *Chlorella*. When the gas-bag is full, excess methane can be burned for heat and power, and the carbon dioxide of combustion again returned to the algal culture. The crew eat the algae — experiments on making palatable foods from *Chlorella* have already been conducted by several laboratories — and return their wastes to the culture and the fermenters.

The whole thing is an ideal hippie retreat. It's a wandering, rent-free, self-contained home in the sky. It can cruise above the clouds in endless sunshine. It can easily dock and undock with other eco-dirigibles. It's the perfect symbol of irresponsible, ecologically virtuous, alternative living. It never has to land. Indeed, so vast is the atmosphere that Daedalus wonders whether the whole increase in the world's population for the next 50 years might be accommodated in a great fleet of eco-dirigibles: both relieving pressure on the Earth's resources and maintaining an alternative, aerial, literally flower-powered culture aloft to counterbalance the rat-racers below.

(*New Scientist*, 23 October 1975)

From Daedalus's notebook

Balloon design. Since the dirigible is primarily a dwelling and not a means of transport, it needn't be streamlined or shaped. So heavy internal bracing isn't required; a completely flexible skin held about spherical by internal pressure will do. An envelope of $r = 50$ m radius (say) will have a volume $V = 4\pi r^3/3 = 5.2 \times 10^5$ m^3. Since methane has molecular weight 16 compared to 29 for air, the available lift will be $29 - 16 = 13$ grams-force per mole or 0.54 kgf per m^3: about 280 tons for the whole balloon.

Let's allow about half this, say 1.5×10^5 kg, for the photosynthetic envelope. Its surface area is $A = 4\pi r^2 = 31\,000$ m^2. If it is essentially a thin layer of aqueous *Chlorella* culture (density 1000 kg m^{-3}) between plastic film its thickness can be $x = 1.5 \times 10^5/(1000 \times 31\,000) = 0.005$ m i.e. 5 mm, so light making a complete pass through the balloon will traverse two walls and at least 10 mm of culture. H. W. Milner ('Algae as Food', *Scientific American*, October 1953, p. 31) quotes $7 - 17$ mm thickness as suitable for productive cultures. A 'quilted' structure for the plastic-film envelope would ensure turbulent flow in the culture, which is important for optimal results. Apart from the envelope, we have about 130 tons of lift available for payload: gondola, fermenters, people, etc.

Energetics. Photosynthesis is:

$$6CO_2 + 5H_2O \overset{hv}{\to} [C_6H_{10}O_5] + 6O_2$$
$$\text{cellulose unit}$$

$$\Delta H = + 2.9 \text{ MJ/unit}$$

Fermentation of cellulose is:

$$[C_6H_{10}O_5] (0.162 \text{ kg}) + H_2O (0.018 \text{ kg}) \to$$
$$3CH_4(0.048 \text{ kg}) + 3CO_2(0.132 \text{ kg})$$

Photosynthesis in *Chlorella* can be as efficient as 8% in bright daylight, which is why the stuff is being studied for food-production and life-support in spacecraft (I. Zelitch, *Photosynthesis, Photorespiration and Plant Productivity*, Academic Press, 1971, p. 275). But let's assume half that in practice. The balloon exposes its cross-section of $\pi r^2 = 7800$ m^2 to the sunlight; if it cruises above the clouds during the day to intercept full sunlight of about 1 kW m^{-2} its energy input is about $P = 8$ MW. At 4% photosynthetic efficiency, this implies production of cellulose at a rate

$M = P \times 0.04 \times 0.162/\Delta H = 0.018\,\text{kg s}^{-1}$, which could be fermented to $\dot{m} = M \times 0.048/0.162 = 0.0053\,\text{kg s}^{-1}$ of methane: about 500 kg of cellulose or 150 kg of methane per 8-hour sunlit day. The 'cellulose' figure in practice will be cellulose/glucose/protein edible algal biomass. A 10-person commune in the balloon might use 20 kg of this daily as a vegetarian diet, leaving 480 kg to be fermented daily to $m = 140\,\text{kg}$ of methane. The heat of combustion of methane is $H = 56\,\text{MJ kg}^{-1}$; so if burnt over 24 hours ($t = 86\,400\,\text{s}$) it could yield power $P = mH/t = 90\,\text{kW}$ continuously for heat/power/propulsion. All these figures could if necessary be multiplied by 2 or 3 using outrigger reflectors to concentrate more sunlight on the balloon; so the basics seem sound. A snag — at 150 kg ($= 210\,\text{m}^3$) a day, it would take about 7 years to fill the balloon from scratch with natural, organic, ecologically virtuous methane. Would a starting charge of North Sea methane count as a sell-out to the military industrial complex?

ECO-DIRIGIBLE

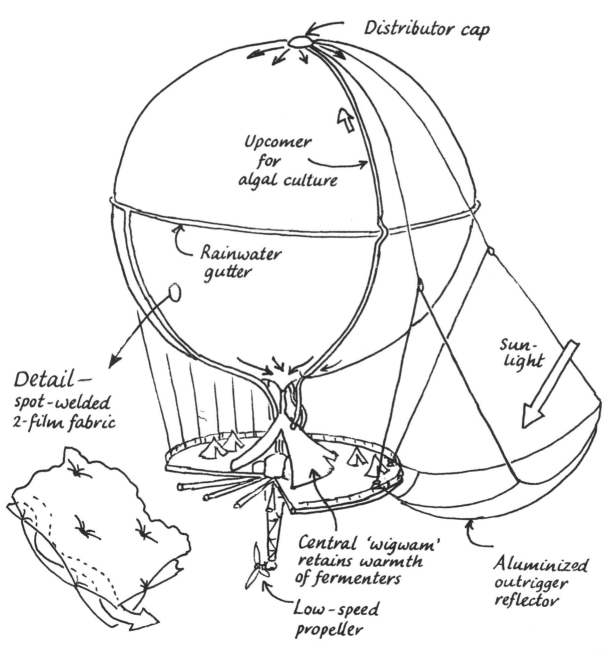

Distributor cap

Upcomer for algal culture

Rainwater gutter

Detail—
spot-welded
2-film fabric

Sun-light

Central 'wigwam'
retains warmth
of fermenters

Low-speed
propeller

Aluminized
outrigger
reflector

Throwing your voice

Daedalus has been musing on the technological applications of smoke-rings. Such vortices, in principle, can transport a fixed sample of gas for any distance; and the bigger they are the better. (A scheme to reduce pollution by making factory-chimneys shoot smoke-rings upwards reached pilot-scale demonstration recently.) So Daedalus is devising a much more efficient vortex-launcher than the conventional pulsed orifice. His machine features a flexible toroidal 'inner tube' full of gas which is pushed down a barrel, spinning about its circular ring-axis as it goes, and opens at the muzzle to release a perfectly pre-formed vortex.

His first idea was to fire huge vortices upwards as ionospheric probes. A hydrogen vortex, for example, would experience steady lift to counter viscous losses. So it should rise indefinitely like a sort of skinless balloon, expanding steadily as it encountered ever lower atmospheric pressures. Stars should show a characteristic twinkle-signature as the vortex 'eclipsed' them with its pattern of refractive-index gradients; the ionosphere should show intriguing and revealing Doppler-shifted reflection changes as it passed through. Ultimately it should rise into the interplanetary gas as a vast bloated tenuous disturbance which, after thousands of years, might finally deliver to some distant planet an authentic sample of earthly gas.

Daedalus's hydrogen-rings would, however, be a severe hazard to aircraft, as they would explode devastatingly if ignited by the jets during an encounter. So he began considering their deliberate use as anti-aircraft weapons. Unfortunately the sedate pace of vortices (a few metres per second) would make them hard to aim accurately at fast-moving targets. Even rings of anaesthetic gas launched at high-flying pigeons or grouse would probably be less effective — though more humane — than conventional shot-guns. As a ballistic weapon, a vortex is more suited to launching gas at fairly static targets on the ground. Its great advantage is that it can't be seen, and gives no clue as to its origin. Used against insurgents or rioters, vortices could carry crude conventional chemicals like tear-gas; but Daedalus is more intrigued by the possibility of using them to create the novel concept of a ballistic smell. An intense odour of burning rubber, say, or maybe steak-and-kidney pudding, or even the dreaded BO, might be more distracting; especially if they seemed to arrive overpoweringly and spontaneously, apparently from the victims themselves. In the confused, urban-insurrection style of future conflicts, the vortex smell-projector may prove a key weapon.

Another useful property of vortices is that they can retain, and even amplify, vibrations. This suggests to Daedalus the intriguing possibility of using them to convey slow sound. His vortex-launching loud-hailer will project a sequence of invisible sonically-vibrating vortices making up a simple message. Travelling slowly, but spreading out hardly at all, they will enigmatically seem to speak from empty air as they envelop their target, minutes after the launcher has made his getaway. As a hit-and-run weapon, the audible equivalent of the graffito-spray, the slow-sound launcher will add a new dimension to public comment. The authorities would presumably fight back by dropping big sonic vortices from aircraft, as a form of mass clandestine public address. Even if 'weighted' with carbon dioxide, such vortices could easily take an hour to descend from 30 000 feet, and nobody would connect them with the original plane even if they remembered it. So they could have a powerful impact on public opinion. The sonic vortex-launcher will be even more effective as a propaganda weapon if its victims, unable to identify the source of the message and perhaps momentarily disoriented by carbon dioxide intoxication, take it to be the voice of conscience.

(*New Scientist*, 13 July 1978)

Daedalus comments

In the 1860s the theory of the vibrations of vortices was developed by the illustrious Helmholtz. It suggested to Sir William Thomson (later Lord Kelvin) the possibility of devising a vortex theory of the atom, to explain the puzzling values of atomic vibration-frequencies then being revealed by spectroscopy. On this theory, atoms were merely vortex-rings in the ether. Since the ether was considered to be utterly without viscosity, the rings were permanent; they moved around, vibrated, and interacted with each other as discrete entities. The theory proved mathematically too intractable to give rise to useful predictions, and atomic spectra remained a puzzle. And it was in an effort to solve this long-standing problem that Niels Bohr, in 1913, took the epoch-making step of applying quantum principles to Rutherford's recently-propounded nuclear model of the atom.

S. E. Widnall and J. P. Sullivan (*Proceedings of the Royal Society A*, Vol. 332, 1973, p. 335) give an intriguing account of an experimental investigation into smoke-rings and their modes of vibration.

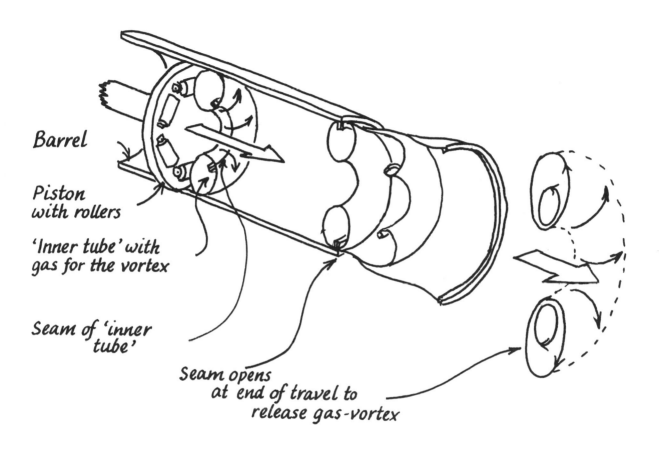

Barrel

Piston
with rollers

'Inner tube' with
gas for the vortex

Seam of 'inner
tube'

Seam opens
at end of travel to
release gas-vortex

Instantaneous shape of
smoke-vortex in a
vibrational mode with
four-fold symmetry

Fire burn and chimney bubble

Atmospheric pollution from factory and power-station chimneys comprises both solid particles and noxious gases like sulphur dioxide and nitrogen oxides. Current practice is merely to discharge the stuff at as high an altitude as possible, so that it comes down widely dispersed and a long way off. Indeed, to help the emissions gain altitude, a factory chimney has recently been invented which blows smoke-rings. Amused by the idea of scaling up a common item of gaseous domestic jollity, Daedalus is now designing a factory chimney which blows enormous smoke-filled soap bubbles. He is confident that by incorporating the new soluble viscoelastic polymers, bubble-mixtures could be made tough enough for this arduous duty. The vast bubbles, essentially liquid-skinned hot-air balloons, would rise many thousands of feet before they burst and discharged their cargo far from ground-level. And by proper attention to bubble-chemistry even this pollution might be prevented. Neutralization of sulphurous smoke with lime or chalk has often been proposed, and it should be a simple matter to incorporate such bases into the bubble-mix as a fine suspension. With a slight excess of lime, the gas-absorbing reactions would go to completion in the long-lasting bubble. The reactive gases would dissolve in the bubble wall, while the solid material formed gradually slid down it, and the suspended ash in the smoke also settled to the floor of the bubble. The ultimate product would be a bubble of clear gas carrying an accreted 'brick' of mixed ash and reaction-product at the bottom, which would fall to earth when at last the bubble burst. The poetic sight of these pearly globes rising from our industrial areas and floating magically away might not compensate for the nuisance of an equal number of bricks descending elsewhere. So a meteorologically-informed anti-bubble squad would shoot them down with pulsed lasers over waste-tips, marshes, or areas in need of land-fill.

(*New Scientist*, 8 March 1973)

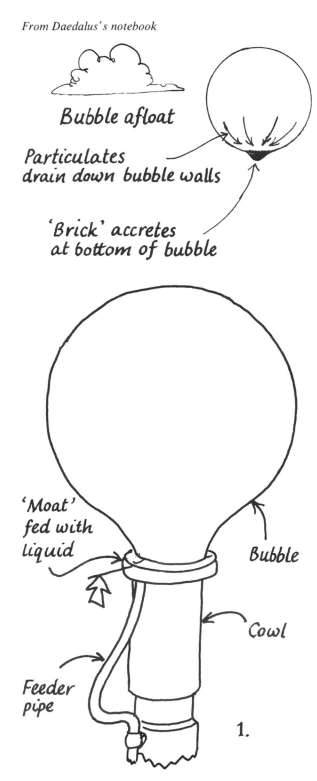

From Daedalus's notebook

Bubble afloat

Particulates drain down bubble walls

'Brick' accretes at bottom of bubble

'Moat' fed with liquid

Bubble

Cowl

Feeder pipe

1.

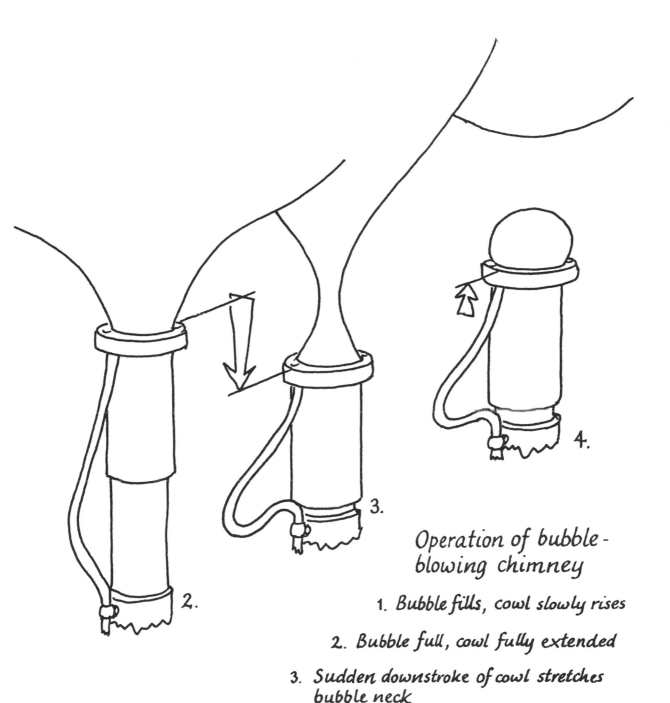

Operation of bubble-
blowing chimney

1. Bubble fills, cowl slowly rises

2. Bubble full, cowl fully extended

3. Sudden downstroke of cowl stretches
   bubble neck

4. Stretched neck is unstable and pinches off,
   releasing full bubble. Cycle repeats

Thicko, égalité, Floppo

The political left is curiously embarrassed by the fact that some people are brighter than others. It constantly seeks to belittle, deny, and suppress the consequences of inherent intelligence differences, and would clearly prefer a world in which we were all equally dim. In this connection Daedalus recalls that the brain is a digital computer. Brain-cells communicate with each other by sequences of identical nerve-pulses, and the intensity of a signal appears to be measured simply by the rate and total number of these pulses. During a pulse, sodium ions are impelled through the porous nerve-sheath into the central cavity, and the nerve cannot transmit another pulse until the biochemical 'sodium pump' mechanism has pumped them out again. Daedalus points out that certain ions, like guanadinium ion, can 'mimic' sodium ions to some extent, and fit in the pores of the nerve-sheath, thus reducing the number of pores available for the diffusion of sodium ions. So DREADCO biochemists are seeking the new elixir of social justice, DREADCO's 'Thicko' (Regd), a guanadinium-derived pharmaceutical which will be absorbed by the nerve-sheaths of the brain and greatly reduce their porosity. The brain's neurones will take longer to recover from each pulse, and their maximum pulse-repetition rate will drop. Ordinary sluggish thought, not using any neurones at their maximum capacity, will be unaffected. But the faster, more efficient brains, belonging to those bright and enterprising souls who unfairly do better than average, will be slowed right down.

Thus the devil of élitism will be painlessly exorcised. Put Thicko in the drinking water and class divisions will collapse. The public schools will atrophy naturally, for nobody will be able to pass (or set) their entrance examinations. The NHS bill will plummet as the neurotic side-effects of over-active intelligence and imagination decline. The over-complexity of society, sustained by the bright few to enslave the dim many, will be swiftly ended.

At first Daedalus was worried that Thicko would sabotage the whole nervous system indiscriminately, reducing strength and sensory acuity as well as intelligence. But he soon realized that Thicko, as a membrane-surface agent, would have its maximum effect on the nerves with proportionally most surface for their volume — i.e. the small-diameter nerves of the brain. By contrast, an agent which acted throughout the volume of a nerve would have most influence on the bigger nerves, i.e. those of the body. So DREADCO pharmacologists are studying the crown-ethers (so called for their bulky ring-like molecules) which bind specifically to sodium ions, slowing their diffusion right down. Diffusion is a true volume-effect, so 'Floppo' (as the new crown-ether pharmaceutical will be called) should limit the carrying capacity only of the body nerves. The two agents together will enable either brain or body to be slowed at will; reducing high intelligence as Thicko does, or decreasing strength and awareness as Floppo will do. Floppo will have many uses. It will not affect the ordinary operations of body nerves, in which successive pulses are separated by enough time for even Floppo-doped tissues to recover fully, but it will cut down their maximum capacity. Thus it will be the ideal antidote to neurotic hyperactivity, preventing the over-eager beaver from straining his body so unmercifully. It will affect the sensory nerves in the same way, reducing only extreme sensations. Thus it will damp the eye's response to harsh glare, or the ear's to deafening sounds: a useful protection nowadays. And as an anaesthetic it could be dosed so as to allow through only tiny amounts of pain, while permitting full sensation up to that point. This would enable a conscious patient to tell the surgeon what he is feeling — valuable feedback in delicate operations. Indeed Floppo may even make possible the do-it-yourself appendicectomy kit.

(*New Scientist*, 23 December 1976 and 6 January 1977)

Daedalus comments

The competition of guanadinium ions with sodium ions in nerves is discussed by D.F. Martin and B.B. Martin in *The Journal of Chemical Education* (Vol. 53 (10), October 1976, p. 614). Saxotoxin, a very dangerous nerve-poison elaborated by the microorganism *Gymnodimium breve*, is a close chemical relative. Some effects of crown-ethers on the transport of alkali metal ions through biological membranes are reported in the *Federation Proceedings* of FASEB (Vol. 27 (6), 1968: D.C. Tosteson, p. 1269; H. Lardy, p. 1278; G. Eisenman, S.M. Ciani and G. Szabo, p. 1289).

Controlled Floppo-doping of the competitors will ensure that all Olympic contests end in a total tie. Much international animosity will thereby be avoided

Seeing the infrared

In a laser, a photon encounters an excited atom, and stimulates it to emit a second, similar photon. These go on to hit other excited atoms, generating more photons, and the result is an uncontrolled avalanche of light. But, says Daedalus, consider an underpumped lasing medium too dilute to 'go critical' in this way. One entering photon might hit an unexcited atom, being absorbed and elevating the atom to its excited state. Another photon might hit this atom, stimulating it to re-emit the 'stored' photon: the two photons would then travel on as a superimposed pair. With a slightly more concentrated lasing medium more intensely pumped, this pair might encounter another excited atom, and thus become a photon-triplet. Overall, about as many photons would leave the system as entered it; but they would come out bunched.

Photon-bunched light will have powerful new properties. For a start, it will look funny. Red bunched light will reflect off red objects in the usual way. But being made of photon-pairs which together have the energy of one blue photon, it will also trigger the eye's blue receptors by two-photon absorption, and will look both red and blue — probably a shifty kind of purple. But it is infrared bunched light that intrigues Daedalus. All ordinary objects emit copious low-energy long-wave IR radiation all the time. So place any object behind the DREADCO photon buncher, clump its photons into aggregates each with the energy of the photons of visible light, and you have free illumination! Multibunched IR would probably look horrid, so a better plan would be to bunch its photons up a bit more into the ultraviolet range of energies. Directed at a standard UV-phosphor of the type used in fluorescent tube lamps, they would excite it by multi-photon absorption. It would then fluoresce to give safe, genuine, visible light. This elegant device upshifts useless ambient IR radiation into useful visible light. In fact it is a 'light-pump', equivalent in principle to the well-established heat-pump which takes low-grade ambient heat and pumps it up to a useful temperature. The laws of thermodynamics enable both devices to deliver far more useful energy, as light or heat respectively, than is needed to drive them.

(*New Scientist*, 26 June 1980)

From Daedalus's notebook

Let's have a laser system with N_1 atoms in the ground state and N_2 atoms of higher energy E. The lasing frequency is then $v = E/h$, and if this frequency has energy-density ρ_v in the system, the excitation-rate $N_1 \rightarrow N_2$ will be $BN_1\rho_v$ (B = Einstein transition probability). Stimulated emission rate $N_2 \rightarrow N_1$ will likewise be $BN_2\rho_v$. Consider n photons entering the system. For each one, its chance of being absorbed via a $1 \rightarrow 2$ transition will be proportional to $BN_1\rho_v$; call it KN_1. So the most probable number of photons lost by absorption will be (for small KN_1) nKN_1. This leaves $n(1 - KN_1)$ photons traversing the system. Each of these has a chance KN_2 of stimulating the emission of a companion photon by a $2 \rightarrow 1$ transition, creating a photon-pair. So the most likely number of such pairs produced will be $n(KN_2)(1 - KN_1)$. We've put n photons in, and got $n(KN_2)(1 - KN_1)$ photon-pairs out. So the bunching-efficiency of the laser is $2KN_2(1 - KN_1)$. This has its maximum possible value when $N_2 = N_1$,

Ordinary uncorrelated light. Chance of two-photon effects (eg. absorption) very low.

Bunched light. Chance of two-photon effects now very high.

i.e. when the external pumping radiation, kicking atoms upstairs by $N_1 \rightarrow N_3 \rightarrow N_2$ transitions, is just failing to invert the population, so the system is just too under-pumped to lase spontaneously. And this maximum value is 0.5 (for $KN_1 = KN_2 = 0.5$). So we might hope to bunch about half the incoming photons. In practice we'll get triplets and so on as well. Even so, 50% efficiency or thereabouts looks attractive.

What will photon-pairs be like? They should undergo physical processes (transmission, scattering, etc.) like their constituent photons; but in chemical processes (absorption, etc.) they'll be more likely to undergo two-photon absorption and behave like a single photon of doubled frequency. Hence, maybe, photon-bunched IR street-lamps which penetrate fog like IR but affect the eye like visible light. And how about a photon-bunching 'anti-sunshade' which takes ordinary dull overcast light and bunches it up to UV for tanning? And since a photon bunch has the same direction and phase as the original photon, you should be able to see an IR image by viewing it through bunching goggles. It's a novel direct-vision thermograph.

Daedalus receives a letter

Myron L. Wolbarsht, PhD
Professor of Opthalmology and
Biomedical Engineering,
Duke University Medical Center,
Durham, N. Carolina, USA
July 23rd 1980

Dear Ariadne
Your friend Daedalus considered (p. 448, June 26, 1980) using photon-bunched light to trigger blue receptors in the eye by two-photon absorption and even considered using low-energy, long-wave infrared radiation bunched together to produce visible light. I enclose a reprint of some of my earlier work which shows that that is indeed possible ('Visual sensitivity of the eye to infrared laser radiation', D. H. Sliney, R. T. Wangemann, J. K. Franks and M. L. Wolbarsht, *J. Opt. Soc. Amer.* Vol 66, 1976, p. 339). I do hope that Daedalus keeps up his suggestions, but he should realize that science is moving so fast these days that even a dreamer sometimes lags rather than leads.
                                        Sincerely

(Signed) M. L. Wolbarsht

(For further bunched light on this contest of priorities, see page 138.)

The odour of celibacy

A pheromone is a chemical emitted by one creature to produce a specific response in another. It's a sort of smell-message, and social insects like ants are guided through their entire instinctive lives by such pheromonal promptings from their fellows. And many creatures — insects, rats, and mice, for example — have courtship rituals triggered by special sex-pheromones given off by one partner and smelt by the other. Daedalus sees this as the basis of a novel anti-sex ploy. DREADCO biologists are working on new rat and mouse baits containing pheromone-antagonists or pheromone-modifiers, to make the poor rodents smell wrong and non-sexy to each other. Their mating instincts will never be triggered, and they will die out in baffled celibacy.

The chemistry of the scheme may be very simple. One theory claims that each substance smells as it does because of the precise vibration-frequencies of its molecular skeleton. Now these frequencies depend crucially on the exact masses of all the atoms in the molecule. So replacing some atoms by other isotope atoms — chemically equivalent but heavier or lighter than the originals — will alter the vibration-frequencies and change the smell. A US nuclear laboratory is said to maintain a 'heavy dog' — fed with heavy water and heavy food, and with the heavy isotope deuterium replacing normal hydrogen throughout its whole body. Daedalus is asking the Americans whether it smells attractive to normal bitches. If it doesn't, then isotopic dog-biscuits could be sold for feeding to bitches on heat. Their come-hither smell would be subtly sabotaged, and the local dogs would completely ignore them. Similarly, isotopically-labelled bait would make an ideal rodent de-sexer. But if isotopic substitution doesn't work, more complex pheromone-antagonist chemistry will have to be devised.

Some DREADCO biologists think it might be easier (and kinder) to make rats smell attractive to others of their own sex, or perhaps to mice. Thus unfruitful but still satisfying homosexual, lesbian, or interspecies relationships will still provide some outlet for their little libidos. Perhaps it is just as well that human sexual 'chemistry' is much more subtle than its rodent analogue, or DREADCO could find itself pioneering a solution to the population problem that would attract almost the ultimate in Papal denunciation!

(*New Scientist*, 26 February 1976)

From Daedalus's notebook

A good account of the skeletal-vibration theory of smell is given by R. H. Wright and J. M. Brand in *Nature* (Vol. 239, 1972, pp. 225 and 226). These workers isolated the alarm-pheromone of fire ants, and measured the skeletal frequencies of its molecule by infrared spectroscopy. They then tested the efficiency of other compounds in triggering alarm-behaviour in fire ants, and discovered that the best molecules all had vibration-frequencies in the same parts of the spectrum, e.g. 330 cm^{-1}, 433 cm^{-1} and 484 cm^{-1} ($\pm 5 \text{ cm}^{-1}$; $1 \text{ cm}^{-1} \triangleq 30 \text{ GHz}$).

Unfortunately, replacing hydrogen by deuterium in these organic molecules, while chemically quite easy, would probably have little effect on frequencies as low as this. A more stringent but chemically trickier test of the theory would be to replace the normal carbon-12 atoms in the molecules by the ^{13}C isotope (lowering the affected frequencies by about $\sqrt{(12/13)}$ or $\sim 4\%$) or ^{14}C (lowering affected frequencies by up to 8% but making the pheromone rather radioactive).

Non-isotopic pheromonal sabotage. The pheromones which induce sexual behaviour are likely to be much more specific than ant-alarm pheromones; one specific compound or composition should be effective while chemical close relatives should have no effect. For as two varieties of mouse, say, evolve away from each other into being separate species, there will come a time when mating between them produces sterile or highly unfit forms. At this stage there will be strong evolutionary pressure to develop separate mating rituals which prevent such wasteful cross-mating. So if sex-pheromones are used in mating, I would expect one of the evolving varieties to make a distinctive change in its pheromonal 'come-hither molecule' at this stage. The more specific the pheromone used by the creatures, the smaller will this change need to be. The two varieties will then have distinct but still very similar sex-pheromones, and each will still be chemically well equipped to generate the pheromone used by the other. So there's a good chance that relatively little chemical interference might resurrect the old pheromones, re-establishing sexual relations between the two species. A pheromone-modifying bait to produce mating between related but mutually sterile rodents, e.g. rats and mice, seems quite on the cards.

Strange-sounding brass

Daedalus recalls with mixed feelings his attempted sabotage of an unswervingly dull organ-recital by secretly dropping solid carbon dioxide into the instrument's wind chest; an attempt foiled, in the event, by a suspicious verger. The theory was that the increased density of the cold CO_2-laden gas reaching the pipes would lower their pitch by a large but fluctuating amount, imparting the diverting impression of a hand-wound gramophone unsteadily running down. In maturer recompense for this ill-fated escapade he now considers how the principle can be put to the service of the melodic muse. He first saw it as a simple way of extending the range of the human voice. A gas-pipe unobtrusively incorporated into her music stand would enable the diva to soar from deepest bass to effortless falsetto by flooding her environment with a gas-mixture appropriate to the aria being performed. Helium and the fluorocarbon gases cover a wide range of densities and are physiologically harmless, but the likelihood of atmospheric turbulence and draughts producing distressing hiccoughs in the vocal performance lead Daedalus instead to musical instrument design. His pyknophone is a flute-like device fed from a number of gas-cylinders by trumpet-style valves. Air is excluded by the outward gas-flow, and the wide variety of usable vapours opens up a whole new range of pitches and timbres for the musician.

Even more attractive is the combination of the instrument with gas-phase chromatography, that technique of separating vapours by passing them down a tube packed with an absorbent which delays them by different times. A pyknophone on the outlet would automatically play a tune as the components emerged in sequence. Only chemical ingenuity limits the musical elaboration attainable by this delightful counterpart of the old-fashioned musical box. Harmonious Gas-Blenders, Ltd, could market appropriate mixtures labelled *The Minute Waltz, Colonel Bogey, Home Sweet Home*, etc., reproducible on the home chromatopyknophone simply by injecting a sample of this genuine 'canned music'.

(*New Scientist*, 18 April 1968)

Daedalus comments

The chromatopyknophone turned out to be less novel than I thought at the time. Gas-chromatography detectors which monitored the changing pitch of an ultrasonic whistle in the exit gas were described in the early 1960s. But my suggestions for applying pyknophonics to music have since been taken up. In the 1970s helium gas was used in at least one concert to give synthetic falsetto to a singer's voice. In 1980 D. A. Davenport *et al.* (*Chemical Technology*, December 1980, p. 774) described a pyknophone, together with a sequence of gases enabling it to sound the notes of the standard musical scale.

During a mild attack of hysteria brought on by an American newscaster referring to 'A change in Mr Nixon's pasture', Daedalus recognized the need for a machine, not to translate between different languages, but merely between different dialects of the same one. Even this relatively simple task suggests microphones and massed circuitry, splitting things into components and feeding them to delay-lines, computer pattern-matching and all the rest of the standard overkill of modern electronics. But Daedalus recalls how strange one's voice sounds on tape, compared to the accent one usually hears through the resonant cavities of one's own skull. It seems to him that the major differences between English accents derive from such changes of pitch emphasis, together with alterations in the duration of component sounds. He therefore felt challenged to design a purely acoustic dialect-transformer. It is a system of speaking tubes and resonant chambers, by which incoming frequencies can be selectively delayed, augmented, echoed or cut short. Non-linear transformations will be handled by reeds, to generate harmonics, and fluid-logic elements and amplifiers fed from a separate gas-supply. Experiment should in due course arrive at the perfect universal 'linguaphone'. This instrument, made preferably in shining brass, will look like a trombonist's nightmare. In addition to its input and output bells it will have a great array of taps, resonators and U-pieces whose adjustment sets it for the pair of dialects in hand. It will be ideal for two-way conversation, for if Scots spoken into one bell emerges as Texan from the other, the reverse must also hold. Thus the great ignored communication barriers of our time will be breached. Australia will at last be in real communication with Pakistan, and Wales with Jamaica. Even the thick, strangled tones of British Rail public address systems may be rescued for rational interpretation.

(*New Scientist*, 10 October 1968)

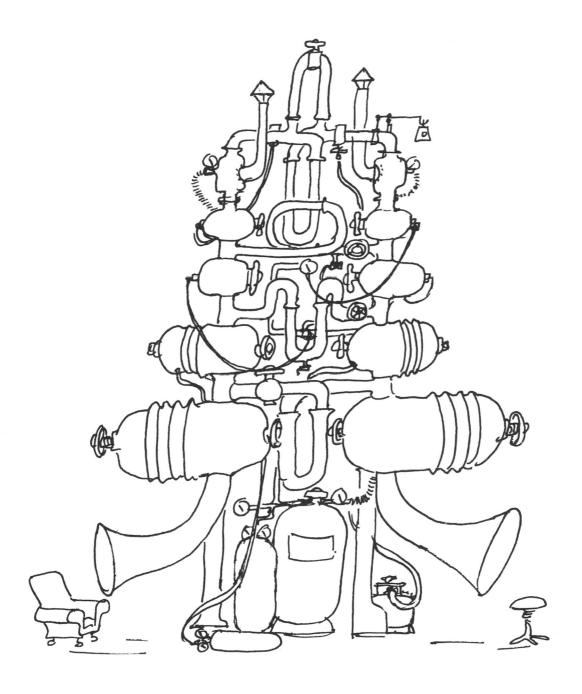

The linguaphone

Global teledisruption

China's Great Leap Forwards is said to have inspired a novel military strategy: if all the Chinese jumped at once they could start an earthquake. Fortunately even a concerted Great Leap Downwards by 700 million Chinese would only release the energy of about 10 tons of TNT, so in its elementary form the scheme would hardly work. But, says Daedalus, suppose the leaps were not simultaneous, but carefully staggered over China by a preset programme? The system would then be the precise reverse of a seismic array detecting a nuclear blast. Normally, the shock-front of the explosion radiates out through the Earth and reaches each geophone of the array at a different time. If one imagines the whole thing played backwards, with each geophone *emitting* a pulse at the right time, the pulses would coalesce into a convergent wave focusing to a shock at the site of the bomb. So Daedalus reckons that the Chinese, purely by coordinated athletics, can deliver the equivalent of a 10-ton bomb to any point on the Earth's surface — the Houses of Parliament, say. Accuracy of this order is well within the capacity of a continent-sized array, although considerable collective skill would be needed to achieve the required microsecond timing-accuracy at each element of the array.

The Western version of this fiendish weapon would presumably use millions of concrete blocks suspended from gantries all over America, and released in precise sequence by an atomic-clock-controlled master computer. Daedalus is inclined to applaud this military advance. A weapon so dispersed could not be crippled by any pre-emptive strike, and could survive the loss of many elements. And its ability to pick off small single targets would put a merciful multiplicity of tiny steps in the escalation-ladder to World War III. It is also ideal for acts of clandestine sabotage, and Daedalus now suspects that the mysterious collapse of box-girder bridges around the globe is a Chinese seismic plot to discredit Western engineering.

(*New Scientist*, 17 May 1973)

Musing on the roundness of the Earth recently, Daedalus decided that it must act as a 'whispering gallery' for sound. Any sound radiating out from its source must travel all round the atmosphere to converge to the antipodal point on the far side of the globe. We don't overhear conversations in New Zealand only because the atmosphere has height. Sound from a point on the ground spreads spherically, up as well as out, and the wavefront fails to converge perfectly on the antipodal point. But a long thin loudspeaker, stretching vertically the height of the atmosphere and launching the sound radially outwards from its whole length, could 'talk' to a similar microphone properly positioned on the far side of the globe. The time-delay would only be 16 hours, and secrecy well assured. Except very near the transmitter and receiver, the sound-energy would be spread over too large a volume of air to be detected. Daedalus has rather a dramatic experiment to test the idea. Tether a high-altitude balloon to Earth with a long length of detonating 'instantaneous fuse'. On firing the fuse, it will all explode practically simultaneously, launching an atmospheric shock which 16 hours later should create a mysterious thunderclap or worse high-pressure catastrophe at the antipodal point. This principle lends itself readily to guerilla activity. How easy to baffle one's antipodal enemies by inexplicable atmospheric explosions sabotaging factories, terrorizing political meetings and collapsing pylons and tower-blocks! (Such vertical structures would of course be particularly vulnerable.) Luckily, until someone invents a hydrogen bomb an inch across and 10 miles high, this form of warfare will be restricted to conventional explosives only. And Daedalus would like to exploit his recently described 'implosives' in this way too. A long vertical implosion in the atmosphere would launch a wave of rarefaction which, travelling all round the globe, would converge to create a sudden vacuum at the antipodal point. Buildings would burst spontaneously and water would enigmatically boil. Suspicion would probably fall on that small thermodynamic chance of all the air molecules simultaneously leaving the vicinity.

(*New Scientist*, 21 October 1976)

SEISMIC SHOCK THEORY

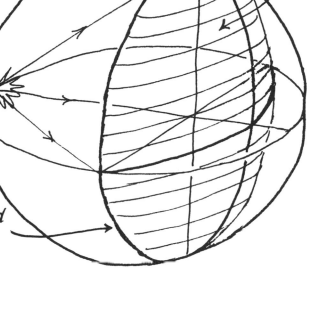

Curved surface
reached by shock
after time t

Source of
shock

Locus of sites from which
simultaneous shocks would
converge on S after time t

Vertical-line
explosion

WHISPERING – GALLERY
THEORY

Concentrically - expanding
sound front

After 16 hours
sound front converges at
antipodal point

Plastic light

This week finds Daedalus musing on new plastics fabrication processes. Many common plastics are made from liquids ('monomers'), which can be polymerized to the solid plastic by exposing them to ultraviolet light, or even visible light. Accordingly a laser-beam shone through a tank of liquid monomer should leave an optically straight fibre in its path. DREADCO physical chemists are therefore experimenting with a tank of liquid monomer surrounded by a complex pattern of mirrors. A laser beam aimed in the right direction zig-zags all around the tank to create an interlaced web of fibres. By proper setting of the mirrors anything from a Brillo-pad to a vest can be made: and with no moving parts at all. This elegant process has many possible extensions. Thus by waggling the laser beam around in the tank, plastic sheeting and tubing of quite complicated forms could be easily made. DREADCO nuclear physicists are inventing a 'plastic-chamber' (by analogy with the cloud-chamber and bubble-chamber) in which the high-energy particles, far more energetic than UV light, leave narrow polymer tracks behind them. Not merely a photograph, but a complete three-dimensional model of their interactions, could then be removed from the tank for study at leisure.

But Daedalus, with an eye on profitable production, is taking the principle still further. He is devising polymerization reactions requiring two stages, each catalysed by light of a separate wavelength. Two different laser-beams traversing the tank would then form a solid spot of polymer at their point of intersection. By scanning this point around, any type of solid object at all could be made up: even complex and re-entrant shapes quite impossible to mould. This effortless optical sculpture would revolutionize the plastic arts in all senses. Designers would be liberated from heavy, expensive steel moulds, and could try out their fancies at will in the laser-bath. Under programmed numerical control, the beams could reproduce any number of identical objects, from string vests 'knitted' by the racing point of beam-intersection to solid garden gnomes whose volume has to be built up by dense scanning. This silent, one-step, infinitely flexible mass-production could even create composite structures, e.g. by lowering a pair of lenses into the bath to have spectacle frames formed around them by the flickering beams. The whole process is in fact a sort of joyful three-dimensional doodling. A pity it would have to be done in the dark, or under harsh monochromatic light, to stop the whole thing setting up solid.

(*New Scientist*, 3 October 1974)

DREADCO receives a Notice of Complaint for Patent Infringement

Shortly after the above scheme was published in *New Scientist*, DREADCO received a Notice of Complaint. It originated from Wyn Kelly Swainson and the Formigraphic Engine Company of California, and drew attention to the following claim, in British Patent 1 240 043 granted to Mr Swainson and issued on 18 August 1971:

CLAIM 2. A method of producing a permanent three-dimensional figure, comprising providing a three-dimensional volume of a radiation-sensitive medium, directing into the said volume at least two beams of different radiations, to which the medium is sensitive, the paths of said beams intersecting within the said volume to define an active region, the medium comprising for responding to each radiation a respective radiation-sensitive active system selectively sensitive to the said radiation and such that individually stimulated systems are incapable of figure formation but as combined at the active region are so capable, and moving the paths of the beams in the said volume, the movements of the individual beams being interrelated so that the active region is moved through the said volume to form in the said material a volume affected by the radiation and having the shape of the desired figure, the last-mentioned volume forming, or the said medium being processed to form from the same volume, the said permanent figure or a mould thereof.

The Notice of Complaint went on to protest that 'since the date of issuance of said Letters Patent, defendant has advocated the use of articles or products responding to and coming within the scope of the disclosure and claims of the said British Letters Patent in infringement of the claims of the said Letters Patent'. In the face of various legal threats contained in the Notice, DREADCO withdrew and apologized cravenly.

Mr Swainson thoughtfully provided photographs of objects made by the process, and in later communications gave details of the progress of his idea. By 1978 a patent remarkably similar to Daedalus's numerical-control notion had been granted to OMTEC Replication (US Patent 4 078 229), from whose press-release of 27 March this diagram is taken:

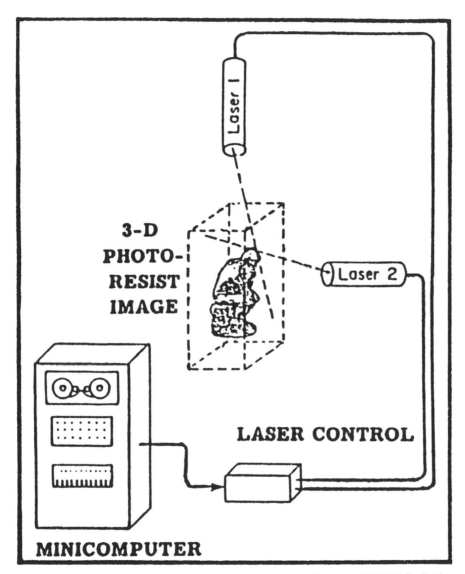

3-D PHOTO-RESIST IMAGE

Laser 1

Laser 2

LASER CONTROL

MINICOMPUTER

Courtesy of OMTEC Replication

91

Radioactive levitation

It is now becoming clear that the fundamental problem of nuclear technology is the disposal of radioactive waste. Alone among the rubbish of civilization, this alien and implacably dangerous stuff cannot be recycled, but must be stored for centuries in (hopefully) reliable seclusion. One recent suggestion, truly worthy of DREADCO, is that rubbish be sucked down into the Earth's interior along the downflow lines of the crustal convection currents. Daedalus now advocates the opposite and seemingly even more technomaniacal course of firing radioactive junk into outer space. So he is inventing the radioactive rocket. He points out that many heavy radioactive nuclei decay by emitting an alpha-particle, a process which produces considerable recoil of the emitting nucleus. If all the alphas were emitted in the same direction, steady thrust would be produced.

At first Daedalus contemplated putting the nuclei into a magnetic field to align them all, but then realized that simply sticking an alpha-shield on one side of the radioactive chunk, permitting the assembly to radiate in one direction only, would do the trick. He calculates that an alpha-emitter of half-life much less than a day should be able to lift itself against gravity and escape into space! Elements of even shorter half-life could lift more than their own weight, so DREADCO reactor experts are seeking operating regimes giving enough of these transient 'hot' residues to lift the rest. Individual rockets need only weigh a few grams or kilograms each—the calculation is independent of absolute mass — and would lift silently away from the reactor with their beastly burden. For neatness' sake Daedalus thinks they should be steered into the Sun. The rapid absorption of alphas by the air would give a brilliant glow, and prevent irradiation of the area in line with this fierce 'exhaust'. Even so, adaptation of the scheme to provide propulsion for ground-based transport is not recommended.

(*New Scientist*, 7 January 1971)

From Daedalus's notebook

Consider a mass M kg of radioisotope, molar mass A. It has $N = ML/A$ atoms in it, where L is Avogadro's constant. If its half-life is $\tau_{1/2}$ seconds, the number of disintegrations per atom per second is $\ln 2/\tau_{1/2}$, i.e. for the whole mass M the number of disintegrations per second is:

$$n = N \ln 2/\tau_{1/2} = \ln 2\, ML/A\tau_{1/2}$$

Each disintegration releases an alpha-particle of mass m and energy $E = \frac{1}{2}mv^2$, and hence of momentum $mv = \sqrt{2Em}$. With simple shielding we can arrange for about a sixth of this to be directed downwards, about a third to be wasted radially outwards, but none to be emitted upwards (Diag. 1). The resulting thrust will equal the downward momentum-flux due to the emitted alphas:

$$\dot{m}v = n\sqrt{2Em}/6$$

thus accelerating the mass M at a rate:

$$a = \dot{m}v/M$$
$$= n\sqrt{2Em}/6M$$
$$= \ln 2\, L\sqrt{2Em}/6A\tau_{1/2}$$

(a) *Unshielded*

(b) *Shielded*

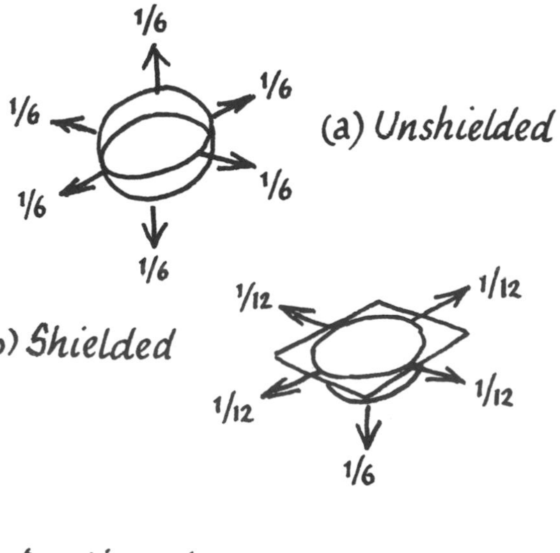

1. Radiated momentum distributions

Since $L = 6.02 \times 10^{23}$ mol^{-1}, and each alpha-particle has mass $m = 6.67 \times 10^{-27}$ kg, this reduces to:

$$a = k\sqrt{E}/A\tau_{1/2}$$

where $k = 8.0 \times 10^9$ kg$^{1/2}$ mol^{-1}.

Typical alpha energies are of the order 1 MeV ($E = 1.6 \times 10^{-13}$ J), whereas the acceleration a only has to exceed that of gravity ($a > 10$ m s^{-2}) for the isotope to fly, so this is beginning to look hopeful. Let's try the formula on some fairly energetic isotopes. ^{250}Fm ($A = 0.250$ kg mol^{-1}, $\tau_{1/2} = 1800$ s, $E = 7.43$ MeV) gives $a = 19$ m s^{-2}; ^{248}Es ($A = 0.248$ kg mol^{-1}, $\tau_{1/2} = 1500$ s, $E = 6.87$ MeV) gives $a = 23$ m s^{-2}. So it looks as if a well-chosen isotope will not only be able to fly, but to lift its own weight or more in further radioactive waste.

We might get up to a sixfold increase in thrust by charging the isotope shield a few MV positive, so that the positively charged alphas are all repelled downwards, instead of just a sixth of them. For big loads of waste the extra complexity might be worth it, and much longer-lived isotopes could fly.

Steering and guidance. The obvious thing is to have servos moving flaps on the shielding to intercept more or less of the sideways radiation, thus angling the total thrust (Diag. 2). Guidance should be ground-based as

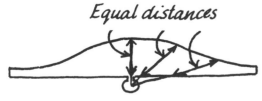

2. Steering system

Resultant thrust

Alpha-emitter

far as possible, with the absolute minimum of on-board gear to be lifted, to save weight. And what form of shielding intercepts the most radiation with the minimum weight of metal? A flat plate could be thinned at the edges to allow for the angled path of the alphas into it (Diag. 3). Or we might use a logarithmic-spiral

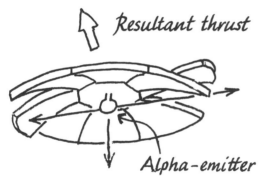

Equal distances

3. Equilateral isotope shield

envelope so that all the alphas hit at the same high angle (Diag. 4). This has the added advantage that the enclosed air adds shielding while the rocket is in the Earth's atmosphere.

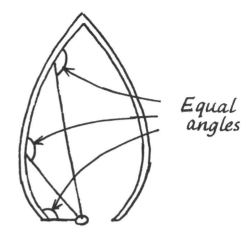

Equal angles

4. Equiangular isotope shield

Trajectory strategy. Unlike conventional rockets, a radioactive rocket exerts a continuous thrust, but one which decays exponentially with time. However, as the rocket lifts away from the Earth, the gravitational force pulling it back falls off as the square of the distance, so the need for thrust decays too. The exact balance of these factors required to get the rocket safely away poses rather a pretty problem. It turns out that there is a very fine balance between the rocket falling back to Earth, or escaping with quite a high final velocity. This is useful, as it means that an escaped rocket will always have surplus velocity to be used in, e.g., steering into the Sun. Thus for an isotope of $1\frac{1}{2}$ hours half-life, an initial thrust just 7.6% in excess of its initial weight will get it away, when it will reach a deep-space speed of some 30 km s^{-1}.

Daedalus comments

About a year after this column was published James Schlesinger, chairman of the US Atomic Energy Commission, suggested in all seriousness that radioactive waste should be lifted into space aboard rockets, and steered into the Sun. He was content to advocate the use of conventional chemical rockets, not realizing that the waste might be made to lift itself. The editor of *New Scientist*, unaware that the idea had originated in his own magazine a little over a year previously, denounced it in high style as 'a load of nuclear crap' (*New Scientist*, 10 February 1972, p. 307). The prophet is ever without honour in his own country, it seems. But the idea didn't go away. In 1977 R. W. Nicholls suggested it in a letter to *Nature* (Vol. 269, p. 556) and on 10 January 1980 *New Scientist* carried a report of an AAAS meeting in San Francisco, in which S. G. Rosen advocated rocket disposal of radioactive waste. Perhaps I should invent the lead umbrella.

The Indian rope-trick and the space-rocket

Conventional chemical rockets are very clumsy, for most of their initial thrust goes into lifting their own fuel. How much neater and economical to lift the fuel beforehand, and top-up the rocket as it goes by! Musing on these lines, Daedalus devised his 'Indian rope-trick' rocket launcher. He recalled the 'instantaneous fuse' used in multiple-charge blasting, which burns at 8000 metres a second; and set DREADCO chemists devising a graduated fuse whose ignition rate would begin slowly and accelerate at (say) $7g$. A long sample of such fuse could be suspended from a balloon and threaded through a hollow rocket-vehicle to emerge from its Venturi thrust-chamber. On lighting the fuse at the bottom, the flame would accelerate upwards at just the rate needed to carry the vehicle on its way, riding the exploding cable. The lifting balloon would at once cast off; but the cable, falling at only $1g$, would remain in the air long enough to do its job.

There is a snag, however. It might take 5000 tons of thick fuse to put 100 tons of vehicle into orbit; how to suspend such a long and heavy Indian rope? Daedalus's brilliant solution is to use the finest known rocket fuel, hydrogen and oxygen, mixed as gases in a long vertical buoyant sausage-balloon. He calculates that a balloon of only 1 m radius could deliver 1000 tons of thrust to a vehicle riding it. For the initial low-velocity stages, the two gases would be safely segregated in the balloon by a partition of graduated fuse, which would burn upwards at the proper acceleration and permit the hydrogen and oxygen to mix and burn after it. At speeds greater than the free flame-rate in hydrogen–oxygen mixture of some 3000 m per second, no partition would be needed. Instead a laser-beam igniter on the rocket would fire the mixture as it sped past. Daedalus is challenged by the idea of using the scheme backwards, with a flame lit at the top and decelerating downwards, as a reverse-thrust decelerator to intercept re-entering rockets and bring them safely to rest: a true shuttle principle. But the exact quoit-like straddling of the tip of the long balloon would require high navigational accuracy.

(*New Scientist*, 20 September 1973)

From Daedalus's notebook

Suppose we have a balloon of a modest 1 m radius. This has volume $V = \pi r^2 h \simeq 3\,\text{m}^3$ per metre of length; with $2H_2 + O_2$ mixture this will include $2\,\text{m}^3$ of H_2. The density of H_2 at s.t.p. is $0.09\,\text{kg m}^{-3}$, so each metre of balloon-length contains 0.18 kg of H_2. Since heat of combustion of H_2 to steam is $121\,\text{MJ kg}^{-1}$, the energy released per metre will be $E = 121 \times 0.18 = 22\,\text{MJ m}^{-1}$. So a vehicle passing at $v\,\text{m s}^{-1}$ will release $P = 22\,v\,\text{MJ s}^{-1}$ i.e. $22\,v$ MW. This is equivalent to a thrust $F = 22\,v/v = 22$ MN or about 2200 tons of thrust. Wow! Assuming thermal efficiency of 45%, a bit optimistic even for a rocket, and we have something like 1000 tons of thrust, independent of vehicle speed.

Note. Efficiency should exceed that of normal rockets. In a conventional rocket the exhaust velocity is constant with respect to the exhaust nozzle, i.e. drops off in the static frame of reference as the rocket gains speed. But with this system the exhaust velocity is constant with respect to the static fuel-balloon. So from the rocket's point of view, exhaust velocity rises as forward speed increases, so efficiency should rise too.

Problem. At high altitudes where the pressure is lower, the balloon diameter must increase if it is to contain the same mass of gas per metre. So make the rocket into a ramjet running *inside* the balloon and consuming the contents as it goes. Then we can give it an automatically-expanding Venturi which increases in diameter to match the balloon. Even so, the 50 km or so of available atmospheric height is still rather a short distance in which to accelerate a rocket up to the orbital velocity of about $8\,\text{km s}^{-1}$. But since the balloon is weightless we can easily lay it at an angle so as to get a longer run. This also minimizes subsequent change of direction to enter orbit.

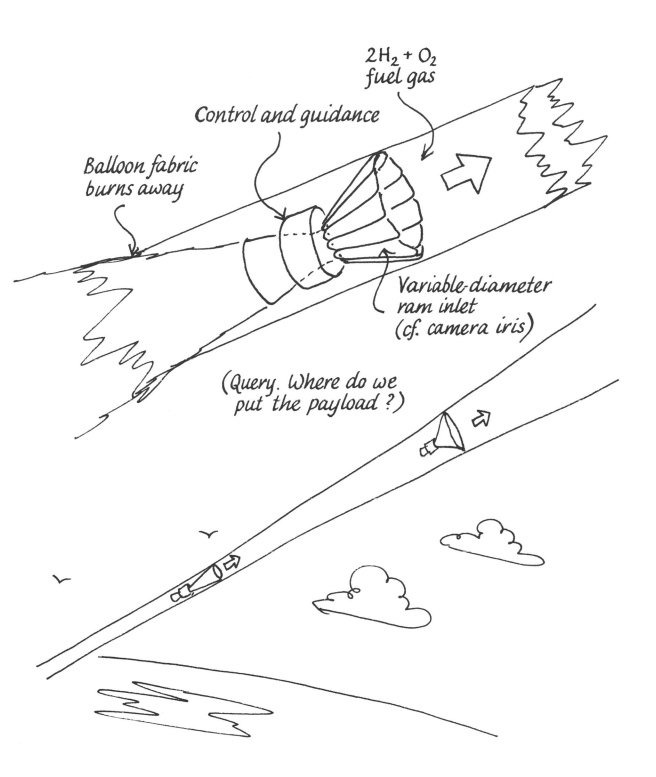

2H$_2$ + O$_2$
fuel gas

Control and guidance

Balloon fabric
burns away

Variable-diameter
ram inlet
(cf. camera iris)

(Query. Where do we
put the payload?)

The thermodynamic way to loveliness

Many women find it extremely hard to lose weight, even on the most drastic of reducing diets. The body's internal 'weight-stat' is very hard to subvert. It reacts by raising the efficiency with which the body extracts nourishment from its food, and completely cancels the effects of the diet. But, muses Daedalus, quite a lot of would-be slimmers could gain greatly in elegance without losing weight at all, simply by a judicious redistribution of their existing fatty tissue. Such a process would not antagonize the weight-stat in any way, and might therefore be quite easy. Now the reason why fatty tissue is important to our appearance is that the body stores it just under the skin, where it acts as a heat-insulator. So some bodily mechanism should exist to deposit it preferentially in the colder areas where it is most needed. Daedalus points out the obvious mechanism: fat (like most substances) is more soluble in liquids at high temperatures than at low ones. Even in a person whose fat-metabolism is in overall equilibrium, fat should be preferentially dissolved from the warm areas, transported in the blood, and deposited in the cold areas. This notion seems to be supported by experience —a recent medical announcement warns that wearers of mini-skirts will tend to get tubby legs!

So Daedalus proposes to optimize the female form by cooling its under-endowed portions and warming those afflicted by ungainly excess. Ladies with the unfortunate 'pear-shaped syndrome' could enhance their charms by wearing DREADCO's electrically warmed 'Hot Pants' in conjunction with the specially cooled 'Frigibra'. Those with over-uniform figures could play both ends against the middle by means of a heated cummerbund and refrigerated bra and bloomers. The temptation merely to expose the underdeveloped areas to the elements may appeal to exhibitionists, but Daedalus feels that climate-independent and thermostatted garments are more becoming as well as more flexible.

Since no total gain or loss of fat is sought, the best figure-shaping garments would be 'adiabatic': all the heat extracted from the areas to be cooled would be pumped back into the areas to be warmed. This would also save the wearer from any overall feeling of being too hot or too cold. The obvious technology to employ is a network of Peltier thermovoltaic junctions, any one of which can either absorb or evolve heat according to the polarity of its connections to its neighbours. Over the complete network, however, half the junctions must necessarily be absorbers, and the other half become emitters. A garment incorporating such a network would then act as a distributed heat-pump, redistributing the wearer's body-heat according to the pattern of the interconnections. The small tempera-

ture-differences it has to pump against would give this heat-pump system a very high 'thermal gain'. Even a modest expenditure of electrical energy from a small battery could pump quite large amounts of heat. DREADCO's 'Adiabatic Combinations' will be studded all over with Peltier junctions, whose connections are set by a program-card inserted into the central controller and power-supply. Thus the areas of local gain or loss can easily be chosen and, by continual updating, can home the wearer unerringly in towards the ravishing ideal of her unspoken dreams. At last! In the privacy of your own home ... Any competent ad-man can take it from there.

(*New Scientist*, 30 September 1971)

From Daedalus's notebook

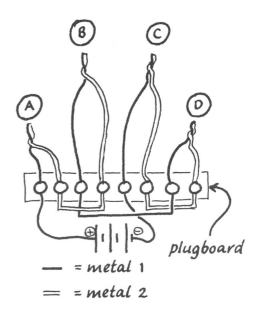

Principle of Adiabatic Combinations —

(showing interconnections required if junctions A and D are to be warmed while B and C are cooled)

plugboard

— = metal 1

= = metal 2

(Junctions with metal 1 +ve evolve heat; those with it −ve absorb heat)

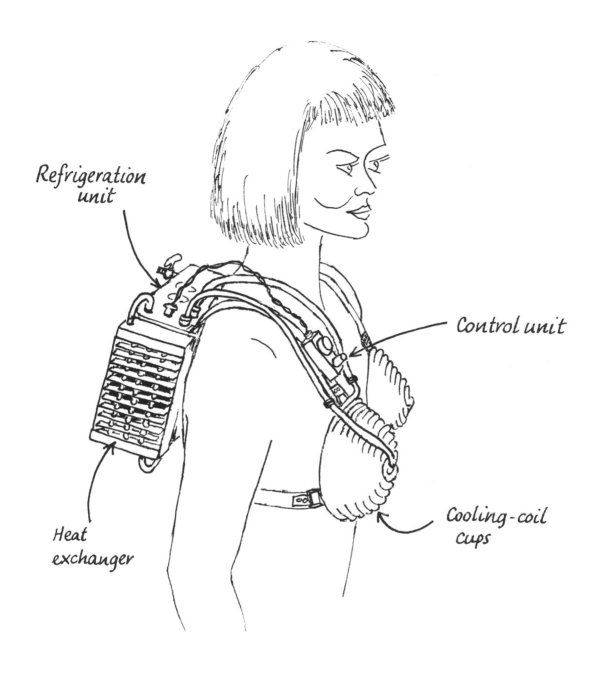

Refrigeration unit

Control unit

Heat exchanger

Cooling-coil cups

DREADCO volunteer wearing experimental laboratory-prototype 'Frigibra'.
(The production model will be cosmetically more acceptable)

Ears to the ground

Daedalus has been pondering the highly resonant nature of the Moon disclosed by recent *Apollo* experiments. Recalling that soldiers marching over a bridge break step to avoid exciting dangerous vibrations in its resonant structure, he hopes that similar precautions will be taken on future Moon walks. Fun-loving scientists have already angered conservationists by advocating the use of nuclear bombs to excite further lunar resonances; DREADCO is studying a still more extreme experiment in which a lunar-surface 'thumper' bangs away steadily at just the right frequency till the Moon — like a resounding glass excited by sustained soprano — shatters! The resulting set of Saturn-type rings around the Earth would make a welcome change in the night sky. Luckily, a terrestrial doomsday-machine on these lines would be foiled by the better damping of the Earth; seismic waves are fairly well attenuated by their passage through it. But such great advances have been made in detecting small seismic waves that Daedalus now proposes a seismic communication system. He envisages a 'transmitter' vibrating against the Earth and a distant gramophone-pickup type of receiver driving amplifier and headphones. Standard AM and FM techniques would provide a multi-plicity of channels, and narrow-band filters would remove the incoherent noise of earthquakes and nuclear tests. Transmission would be completely free from over-the-horizon effects, though really long-range work would probably need some sort of continuously modulated explosion as a transmitter. So Daedalus is concentrating on short-range applications, and has already devised his 'talkie-walkie' communication-boots, whose highly tuned vibrosensitive soles act as transceivers for the wearer's head-set — even ringing a little bell to warn of an incoming call. On the Moon, whose sharp curvature and absence of ionosphere give radio a very short range, the system should be ideal.

(*New Scientist*, 8 January 1970)

From Daedalus's notebook

Isotropic seismic waves launched into the Earth would spread in volume and be attenuated as r^2. Rayleigh surface waves, which propagate like sea-waves along a solid surface, spread only superficially and are atten-uated only as r; they seem a much better bet. (The Indian with ear-to-ground detecting the approaching cavalry must be intercepting mainly Rayleigh waves.) They travel at about 90% of the velocity of shear waves: say 3 km s^{-1} for granite, so propagation delays shouldn't be too troublesome. To excite them, the vibrating sole should oscillate in some oblique direction rather than vertically. One advantage of using boots as transmitters: the mass and compliance of the wearer helps to direct most of the energy down-wards into the Earth. But as receivers, they may contribute creakings, pulse-noises, etc., from the wearer. Maybe we should have a separate 'shooting-stick' receiver to stick in the ground some way away, sending its signal to the user via a cable? Piezoelectric transducers seem preferable to electromagnetic ones for this work: they're lighter, simpler, more robust, and better adapted to transverse excitation. But we'll need a good battery pack to generate the high voltages needed for transmission.

Daedalus comments

DREADCO was beaten to the draw on this develop-ment! Some while after this column appeared, I came across a paper 'Communication via Seismic Waves employing 80-Hz Resonant Seismic Trans-ducers' in *IEEE Transactions* (Vol. COM-16(3), June 1968, p. 439) by K. Ikrath and W.A. Schneider of the US Army Electronics Command. These workers used electromagnetic transducers of 10 or 200 W capacity, and demonstrated reception over ranges up to 1 km. Their system appeared to excite isotropic seismic disturbances except in some interest-ing experiments on lake ice, where they excited surface flexural modes in the ice which travelled horizontally with little loss. But some such modes couple to the air above the ice. In one experiment on the ice, an air-coupled flexural mode was transmitted. 'The resulting audible bursts, which sounded like ice breaking, made the receiver party flee to shore.' Exciting times at US Army Electronics Command! There must be an anti-Russian weapon in it somewhere . . .

THE SEISMIC COMMUNICATION - BOOT

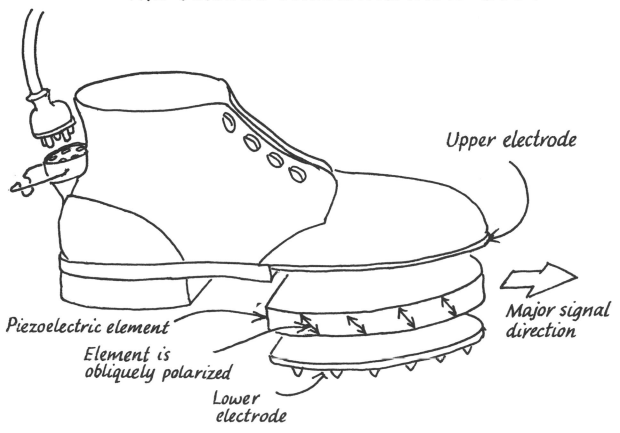

Upper electrode

Major signal direction

Piezoelectric element

Element is obliquely polarized

Lower electrode

Pair of boots as a phased array

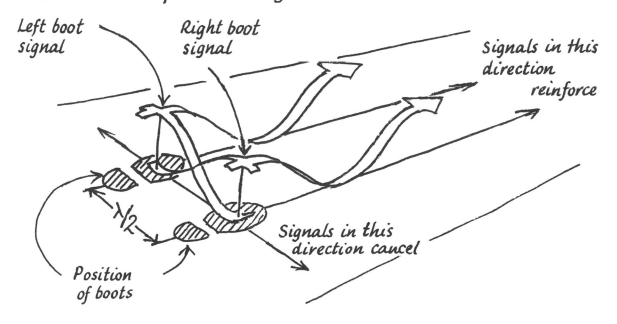

Left boot signal

Right boot signal

Signals in this direction reinforce

$\lambda/2$

Position of boots

Signals in this direction cancel

Rolling along on old man river

Daedalus has been contemplating the new generation of waterborne vehicles: hydrofoils, hovercraft, hydroplanes and so on. He reflects that although these machines experience much less water-friction than floating vessels, this is still an important limit on their speed (except for the hovercraft which pays for its lack of water-contact by having to employ air as the propulsion medium). A typical hydrofoil, like a water-ski, is almost an arc of a circle intersecting the surface: so Daedalus suggests completing the circle to obtain a big wheel. Such a wheel travelling over the water would experience lift by virtue of its advancing profile, but because of its rotation would have practically no drag. His scheme is to build a watercraft mounted on four such drums, floating on them when stationary but lifting up on them to run over the water at speed. Propulsion could be by a conventional screw, but also possibly by powering the wheels and fitting them with small paddles. This would provide a vehicle which, like the hovercraft, could easily run up a beach for discharge. The difficulty is that to travel easily over large waves might necessitate an incongruously large wheel-diameter, possibly resulting in a vehicle consisting of one huge drum, with all the works inside and hanging from the axle. Alternatively a system of tank-like tracks could be adopted, whose profile need not be circular but could be contoured for maximum hydrodynamic efficiency. Daedalus is uncertain whether conventional steering gear would work, or whether differential braking of the wheels or tracks on one side would be necessary. But this type of boat, unlike all others, would be unaffected by the encrustations of barnacles, etc., on its submerged parts, and would in any case tend to throw these off by centrifugal force.

(*New Scientist*, 10 March 1966)

Daedalus comments

Within about a year, both the Americans and the Russians had come up with shameful imitations of this DREADCO development. *New Scientist* published reports of their activities (facing page); and a diagram of the American version (below) which could well have been taken from my own notebook. Still, the patents covering these water-vehicles are presumably rendered invalid by my prior publication.

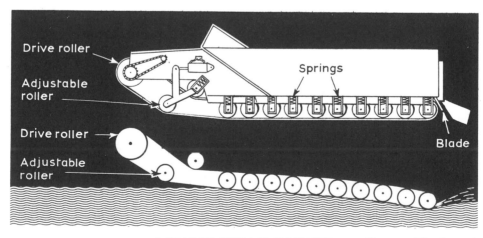

A BOAT TO RUN ACROSS WATER   (Courtesy of *New Scientist*)

New Scientist, 16 June 1966, p. 706:

A BOAT TO RUN ACROSS WATER

In *New Scientist*, in March ('Ariadne', Vol. 29, p. 638), Daedalus was contemplating the new generation of water-borne vehicles and came up with the idea that the best form of transport would be a vessel which could 'run' across the water. A patent described in the United States Navy's Research Reviews, shows that Daedalus was not so far away from a practical suggestion.

The patent, taken out by Berger M. Shepard of the Naval Ordnance Laboratory, White Oak, Maryland, describes a craft which runs over the water on tracks rather like the way in which a tank runs over land. The belt, which provides the craft with its motive force, is as wide as the boat and traverses the boat from stem to stern on rollers which lie across the boat.

When the belt is moving at low speeds — and likewise the craft — the hull lies deep in the water. But as speed is increased, the belt supports the boat at or near the water's surface. The advantage of this system of drive is in mimimizing the effect of friction between the vessel and water when the craft is travelling at high speeds.

In effect, the boat moves forwards while the belt moves backwards in relation to the water's surface.

To secure the best possible thrust and lift at all speeds, while minimizing the drag, the belt system is designed so that it travels through the water almost horizontally. This problem defeated other designers of similar vessels. The new design includes a pair of movable rollers near the bow which permit wide variation of the angle of attack of the boat. The invention also calls for a small enough roller to be mounted at the stern to produce high centrifugal forces to assist in breaking the adhesion between water and belt and also a device to prevent the water from being carried along with the boat.

New Scientist, 13 July 1967, p. 78:

USSR

ROLLING OVER THE MAIN

A new kind of craft, promising to be a competitor of both the air-cushion and hydrofoil ship, is undergoing trials. It uses revolving cylinders set at each side of the hull fore and aft to propel the craft and also to lift it virtually clear of the water and so to reduce drag as to permit relatively high speed. It appears to reach maximum efficiency at 75 knots.

This new application of the Magnus effect was developed by the Soviet engineer Viktor Podorvanov after conducting a series of experiments to determine changes in the lift of rotating cylinders of different diameters and spans at various forward and rotational speeds. In the course of these experiments, which provided exact information related to power output for a given effect, he was impressed by the fact that no bow wave was created because no water is shifted by the hull.

A prototype was built with two big cylinders, set one on each side of the bow, and smaller ones on each side near the stern, all four driven by a main engine through hydraulic couplings. On starting, the cylinders are partly submerged. Their effect is to raise the hull as speed builds up until the craft is practically clear of the water and progresses with a bouncing movement. It has therefore to be provided with a vertical stabilizer and air rudder. The hull itself is designed more on aircraft than ship lines.

Tests are now being made at sea to prove the ability of the cylinder boat to negotiate waves. Figures have already been obtained to relate power to performance and the designer asserts that the energy required is 'much less' than that needed for the propulsion of a comparable hydrofoil vessel. At a speed of 60 knots the cylinders barely touch the water and between bounces the boat is airborne. He claims that the present cruising speed of 75 knots could be doubled or trebled, without causing the craft to lose its bouncing characteristic, by varying cylinder design and speed.

Provided trials prove successful structurally as well as operationally, this type of vessel would become an alternative to the air cushion vehicle as an amphibian. Its cylinders would allow it to climb up a shore or slipway and, if they had a covering of rubber, could be used to travel along a road. They are said, in any event, to be immune from damage by logs or other flotsam because they roll over them.

The design has been patented and rights of exploitation are in the hands of the Soviet foreign trade firm of Litsenzintong.

Aqua-combinations

Daedalus enjoys skin-diving, but feels the technology could be improved. He proposes to counter the main hostile features of the sub-aquatic environment — cold and the lack of breathable air — by a novel invention: his 'aqua-combinations'. They were inspired by those water-repellent garments which are unwetted by liquid water drops while remaining permeable to air. Daedalus calculates that a water-repellent fabric with holes a micrometre or so across would resist ingress of water even under an extra atmosphere of hydraulic pressure. So his aqua-combinations are woven from many layers of 1 μm silicone fibre. They envelop the wearer up to the neck, and resemble wool in appearance and thermal insulation, but water simply cannot get through them (although air and water-vapour from the skin can).

To complete the wearer's protection, a waterproof neck-seal retains a goldfish-bowl style helmet. To provide air, many small tubes inside the garment connect the helmet to all parts of the porous surface of the aqua-combs. Daedalus recalls that a thin silicone-rubber membrane is sufficiently permeable to gases to act as a 'gill', by which dissolved oxygen is desorbed from water, while carbon dioxide is dissolved. He argues that the free water surface within the minute pores of the garment must be even more permeable to these gases, so the whole garment — whose total area of say 2.5 square metres will be multiplied several-fold by the convolutions of the liquid surface against the water-repellent fibres — should act like a huge gill. Like a frog, the wearer will breathe with his whole surface; he will be able to inhabit air and water indifferently, and will even be able to dive to about 10 m before the water-pressure overcomes the water-repellency of his garment and saturates it. Even then a small oxygen cylinder, pressurizing on demand the tiny free volume of his suit, could oppose the pressure and give him the freedom of the depths. But like other denizens of the deep, he would be vulnerable to detergent pollution. This would reduce the surface tension of the water and enable it to wet and penetrate his garment, forcing him cold and gasping to the surface.

(*New Scientist*, 25 November 1976)

From Daedalus's notebook

What diameter pores do we need in a garment to prevent water from entering under 1 atm ($\Delta p = 10^5 \text{N m}^{-2}$) of extra pressure? The surface tension of water is $\gamma = 0.074$ N m^{-1} at 10 °C: so for a pore in a completely water-repellent surface, $r = 2\gamma/\Delta p = 2 \times 0.074/10^5 = 1.5 \times 10^{-6}$ m radius or 3 μm diameter. This means weaving with about 1 μm thread. Microporous silicone film might be easier to handle.

How much surface do we need? The area of the human lungs is about 30 m². It's merely moist surface: oxygen dissolves in its water, and then is transferred by liquid diffusion to the bloodstream. (When our ancestors came out of the sea, they took enough of it with them to breathe through. We're still water-breathing animals really; we just carry our own. Intriguing.) So we might need to feed the lungs with oxygen in back-to-back manner from another $A = 30$ m² of moist surface. Things aren't as bad as this for the lung is at 37 °C, for which oxygen-solubility is only 0.024 vol/vol; whereas at a sea-temperature of say 10 °C the solubility is greater: 0.038 vol/vol. The colder surface, handling more concentrated oxygen solution, can be smaller: $A = 30 \times 0.024/0.038 = 19$ m². Furthermore, oxygen is more soluble in water than nitrogen, so the gas dissolved in water is already enriched with oxygen. The old Mallet's process for enriching air with oxygen got 35% O_2, 63% N_2 from aerated water, compared to 21%, 78% in ordinary air. So since the combs will be delivering this enriched oxygen anyway, they can be smaller still: $A = 19 \times 21/35 = 11$ m². The meniscus bulge of water between the fibres should nearly double the effective surface of the garment anyway, implying finally $A = 5.5$ m² of garment. A typical tailoring requirement for whole-body coverage might be 2–2.5 m², so we only need to find a factor of 2 or 3 by means of corrugations, pleats, wattles, etc. Seems feasible!

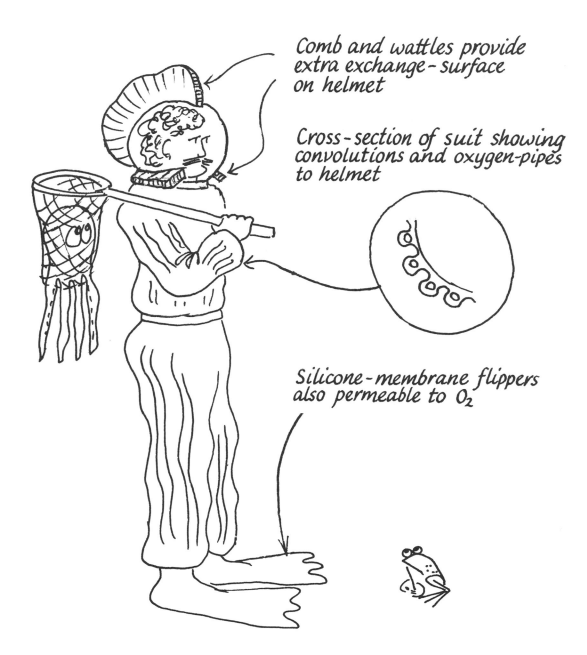

Comb and wattles provide extra exchange-surface on helmet

Cross-section of suit showing convolutions and oxygen-pipes to helmet

Silicone-membrane flippers also permeable to O_2

Thermal glidoons (1)

Modern aviation is dangerously dependent on increasingly scarce hydrocarbon fuels. Even a return to zeppelins would not eliminate this dependence, but only reduce it. And yet, says Daedalus, the atmosphere is itself a vast source of entirely free energy, if we have the wit to exploit it. He is referring to the rapid drop in air-temperature with altitude (at 25 000 feet it's about $-33\,°C$). Ammonia gas, he points out, is lighter than air, but liquefies at $-33\,°C$. Hence, one might think, a balloon filled with ammonia would rise rapidly to 25 000 feet, and then lose lift by the liquefaction of its gas. Unfortunately the reduction of atmospheric pressure with height complicates matters. It steadily lowers the liquefaction-temperature of ammonia with altitude, so that in fact at any height this temperature is always a bit lower than ambient. Daedalus plans to overcome this by putting his ammonia in a somewhat elastic balloon which will always squeeze it to about 0.1 atmospheres greater pressure than the atmosphere outside. This will raise its condensation-point sufficiently for the ammonia to liquefy at about 34 000 feet. With good design, such a balloon would conduct a perpetual oscillation. It would rise rapidly to 34 000 feet, lose lift by liquefaction, and descend through the warmer layers of the lower atmosphere until enough ammonia had reboiled to send it up again.

Now any oscillator must incorporate sufficient gain and overshoot to prevent it settling into some equilibrium condition. For the ammonia balloon, the main such danger is that of becoming 'becalmed' at high altitude in a state of neutral buoyancy and partial condensation. So Daedalus hopes to model his craft's elastic envelope on that of the ordinary toy balloon, which exerts its highest pressure on the gas inside when its diameter is small (that's why it takes a lot of puff to *start* blowing one up). Thus when the ammonia begins to condense and the envelope starts shrinking, the internal pressure — and therefore the condensation-temperature of the ammonia — will rise. So

condensation will continue even as the craft enters its fall into the warmer regions below. The converse problem of delaying the reboil until the craft has dropped almost to earth, is much easier. The liquid ammonia will run to the bottom of the balloon, where appropriately lagged receivers will reduce its subsequent heat-uptake to whatever degree proves necessary.

To convert this paragon of perpetual motion into a commercially attractive craft, it only remains to build it in the form of a glider. Then its descent from altitude becomes not merely precipitate, but useful: for even a modest glider can travel 20 feet forward for every 1 foot of height lost. And when at lower altitude the ammonia begins to reboil and the craft regains its lift, it will be able to 'glide' upwards again towards its maximum height, thus covering some 250 miles per cycle. Such a tubby, inflated glider (or 'glidoon') is best controlled not by conventional hinged control surfaces, but by warping the whole flexible structure — the 'wing-warp' technique that the Wright brothers used to steer their original aeroplane. Additional control will come from varying the lift. At the maximum height of each vertical 'tack', the condensing liquid ammonia will be led into pressure-vessels. At lower altitudes it will be released controllably to raise the buoyancy; to land the craft it will not be released at all, so that the glidoon will come to earth like a normal glider. But to take off subsequently, the valves will be opened and the ammonia will come boiling out to inflate the envelope and lift the craft silently into the air on its next journey.

This beautiful vessel at last makes possible an alternative aviation not based on noisy, smelly, fuel-guzzling technomania. By repeated 'tacking' between low and high altitudes, any distance could be easily covered. Silent, simple, and slow, Daedalus's 'thermal glidoons' will encompass the globe with goods and passengers at almost negligible cost.

(*New Scientist*, 10 February 1972)

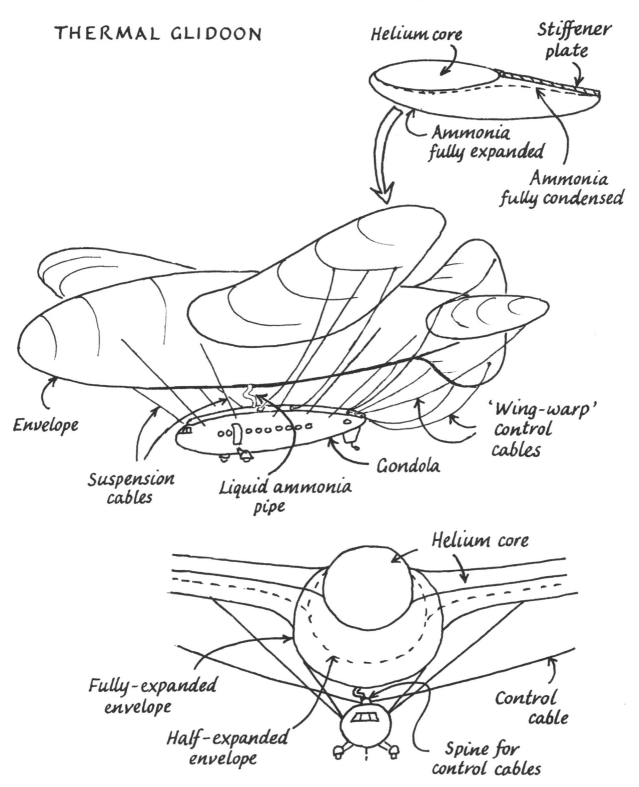

From Daedalus's notebook

Wing cross-section

THERMAL GLIDOON

Helium core

Stiffener plate

Ammonia fully expanded

Ammonia fully condensed

Envelope

Suspension cables

Liquid ammonia pipe

Gondola

'Wing-warp' control cables

Helium core

Fully-expanded envelope

Half-expanded envelope

Spine for control cables

Control cable

Thermal glidoons (2)

From Daedalus's notebook

The temperature of the atmosphere drops from about 15 °C at sea-level to − 57 °C at 11 000 metres. For a thermal glidoon to work, we need a gas lighter than air and condensing in the range (say) − 10 to − 50 °C. Ammonia gas ($NH_3 = 17$), condensing at − 33°C at normal atmospheric pressure, is the only feasible choice.

Unfortunately atmospheric pressure falls off with height. So the condensation temperature of ammonia, which decreases with decreasing pressure, falls too, and is always below the local temperature at any height. Pity. There are three ways out:

(a) Mix the ammonia gas with a little water-vapour or some other solvent vapours to raise the condensation-point of the mixture, or add absorbents like calcium chloride or activated charcoal on which the ammonia may condense at higher temperatures than it would do on its own. I'd have to grope through a lot of woolly multicomponent chemistry in the hope of finding some mix that would work.

(b) Pressurize the envelope of the balloon. In an elastic envelope the pressure inside is always greater than that outside. This raises the condensation-temperature of the gas: also, unfortunately, its density. But with luck we can find an overpressure which makes the ammonia condense at some convenient altitude without the increase of density robbing us of too much lift.

(c) Incorporate a permanent source of lift, e.g. a helium balloon. Then we can overpressure the ammonia even above local air-density without losing all lift. In the extreme case, imagine a helium balloon coupled with a completely inextensible ammonia balloon full of ammonia at atmospheric pressure. As the diagram shows, by the time the assembly had reached 5000 m, the ammonia would have the same density as the surrounding air and would contribute no net lift. But the helium balloon would continue to haul it upwards towards 8000 m (− 37 °C) where the ammonia (which would have lost pressure to 0.82 atm through its cooling) would condense. The sudden collapse of the envelope would so reduce the total buoyancy as to bring the whole assembly down again till the ammonia had reboiled.

A combination of strategies (b) and (c) seems indicated. A bit of playing around with the data (Diag.) suggests that an ammonia balloon overpressured to about 0.1 atm at high altitude is roughly optimal.

What lift would this balloon have? To simplify things, assume a constant 0.1-atm overpressure, and ignore any contribution from helium. Then at sea-level and 15 °C, the density of ammonia is 0.73 kg m^{-3}; over-pressured to 1.1 atm it will be 0.80 kg m^{-3}. The density of sea-level air is 1.23 kg m^{-3}, so the lift available is $(1.23 − 0.80) = 0.43$ kgf per m^3 of ammonia, or 0.54 kgf per kg of ammonia.

At condensation-height, around 10 500 m, the density of air is 0.39 kg m^{-3} at 0.24 atm pressure. The ammonia is therefore at 0.34 atm. Ambient temperature is − 55 °C, and since the condensation-point of ammonia at this pressure is − 53 °C, it will start to condense. Its density under these conditions is 0.32 kg m^{-3}, so its lift is $(0.39 − 0.32) = 0.07$ kgf per m^3 of ammonia, or 0.22 kgf per kg of ammonia. So the lift of the ammonia has about halved in its rise. To reach condensation altitude the structure + payload must not weigh more than about 200 g per kg of ammonia carried (things improve of course when we add helium).

Can we make an elastic envelope for this duty? The overpressure in a balloon of skin-tension γ is $p = 2\gamma/r$; so if the tension γ increases linearly with radius r (i.e. ideal elasticity) the overpressure p is constant for all radii, which is a good start. In a toy balloon, the thickening of the rubber as r decreases tends to keep γ high, so p increases as r drops, which is what we want. There will be complications with variations of elasticity with temperature, but the thing seems feasible. The balloon volume increases 2.5 times during the ascent (ammonia density drops from 0.80 to 0.32 kg m^{-3}) so the linear extension required of the envelope is only $\sqrt[3]{2.5} = 1.36$ i.e. 36%. That's not very extreme.

Construction of thermal glidoons. There isn't nearly as much lift to play with as in a pure helium balloon. So a heavy internally braced structure is out. The Goodyear Company once made an inflated rubber glider, and for an overpressured balloon this structure is almost perfect. It also requires some sort of permanently inflated 'core', or the whole glider will collapse to a floppy mass of wet rubber when the ammonia condenses. A helium-filled 'core glider', with an outer elastic skin containing the ammonia, is ideal. One advantage here: we may be able to arrange the aerofoil section of the fully expanded wing to be ideal for gliding upwards on the rising tacks of glidoon journeys, while that of the core wing is ideal for gliding downwards on the descent tacks. This is neater than e.g. turning the whole craft upside down.

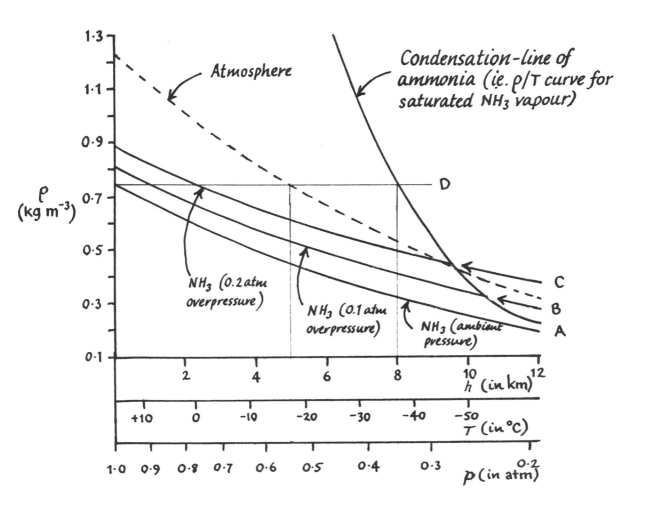

Gas-density ρ as a function of height h for the atmosphere, and for ammonia in thermal equilibrium with it. The corresponding atmospheric temperature T and pressure p are also marked as horizontal scales. Note that

A. Ammonia at ambient pressure is less dense than the atmosphere at all heights, but never reaches the ammonia condensation line

B. Ammonia overpressured by 0.1 atm remains less dense than the atmosphere right up until it crosses the condensation-line and condenses at 10.5 km altitude

C. Ammonia overpressured by 0.2 atm reaches the density of the surrounding atmosphere and thus loses all lift at about 9 km altitude, just before reaching a height at which it would condense

D. A constant-volume balloon containing ammonia at 1 atm would rise till it reached the same density as the surrounding air at 5 km; if hauled further aloft by an auxiliary balloon it would condense at 8 km

The copper apples of the sun

Rio Tinto Zinc's plans to despoil parts of Snowdonia with open-cast copper mines has stirred up considerable controversy (*New Scientist*, **12 November 1970, p. 317). This sort of thing will become more common as our myopic exploitation-economy finds itself with fewer and fewer mineral resources to exploit, and those of decreasing quality (the Snowdonia ore is only 0.5% copper). At this rate we shall soon be desperately overturning hundreds of square miles of land in search of tiny remnants of overlooked ore. In principle, however, the removal of 0.5% of a soil's content need hardly disturb it at all; and Daedalus recalls how many studies of the fate of pesticides and radioactive waste have revealed some organism concentrating a substance thousands of times. So DREADCO agricultural experts are developing special metalliferous trees to replace open-cast mining. They are cultivating suitable starting plants (e.g. green peas, which contain quite a lot of copper already) in radioactive and copper-rich soil, hoping to induce mutations of high extractive efficiency. With luck the element may even become essential to the plant, perhaps replacing the magnesium in some of the chlorophyll and enchancing photosynthesis by capturing additional wavelengths of the solar spectrum.**

Daedalus's ultimate goal is a metal-tree which will send a deep and complex root-system into the soil in search of the metal on which it has been 'hooked'. It should deliver heavy, practically metallic fruit — ideal for windfall harvesting but a nasty surprise for finches and latter-day Newtons. A plantation replacing the present Forestry Commission acreage in Snowdonia National Park could deliver 10^5 tons of copper per square mile per year for only 1 metre per year of vertical root-growth! And fertilization by finely-ground cuprous rubbish (old TV sets, defused ammunition, brass bedsteads, etc.) would ensure leaching of the copper back into the soil for recovery. This would result in effective recycling without any industrialist having openly to mend his ways and start designing his products with ease of reclamation in mind.

(*New Scientist*, 17 December 1970)

Daedalus comments

Quite a number of plants have the ability to extract metals from the soil in which they grow. In many cases these 'hyperaccumulator' plants colonize metal-rich soils which are unfavourable to their competitors. They take up the metal, detoxify it and store it as a harmless and inert burden in their tissues.

One very dramatic example came to light about a year after this column appeared. B. C. Severne and R. R. Brooks of Massey University, New Zealand (*Planta*, Vol. 103, 1972, p. 91) discovered in Western Australia a variety of the shrub *Hybanthus floribundus* which could concentrate nickel in its tissues to an extent of 10% of the dry weight, and to a world vegeto-metallic record of 23% in the leaves! Considering that good nickel ore may contain only 3% of the metal, agricultural mining is clearly an attractive proposition. Certainly this is the opinion of Alois Bumbalek, to whom British Patent 1 481 557 was granted in 1977. His claim is that certain fruit trees — citrus and banana in particular — will if deprived of certain key elements attempt to restore their metabolic balance by taking up others. Thus if deprived of potassium, the trees will accumulate gold as a first option, with silver and lead as secondary choices. Magnesium deficiency leads them to accumulate uranium. Useful amounts of uranium, thorium and titanium have, it seems, been found in Ecuadorian and Honduran bananas from the operation of this mechanism. *New Scientist*, commenting on the matter on 22 December 1977 (p. 771) remarked: 'The method offers the possibility that useful quantities of valuable minerals could be extracted from ground normally viewed as too poor to mine, by the simple expedient of growing trees on the surface and harvesting their fruit for controlled incineration.' So once again, others follow where DREADCO leads!

Many plants can extract trace metals from their local soil, and concentrate them in specific tissues

Hydraulic coal-heaving

Conventional coal-mining is so clumsy and dangerous that Daedalus is very pleased to announce his new and painless mining technology. Coal is a remarkably light mineral. Its density is less than that of dry-cleaning fluid, or even strong calcium chloride solution. So, says Daedalus, fill up a mine with dry-cleaning fluid and the coal will float to the surface up the shafts! You need only sink two shafts into a coal-seam and connect them by a boring; then pump your dense liquid down one shaft, filter off the floating coal as it emerges from the other shaft, and recycle the liquid. Some means of loosening the underground coal will be needed, however. Daedalus points out that coal is full of cracks, and like other organic polymers it is easily weakened by organic solvents. So DREADCO chemists are devising a solvent formulation combining high density, high solvent-action, and good crack-penetrating detergency, so that the turbulence of rapid pumping will cause it to strip coal steadily from the underground seam by the force of its high-pressure passage.

For deeper mines with more solidly compacted coal, Daedalus has another strategy. He recalls that newly won coal oxidizes spontaneously in the air (indeed it can catch fire dangerously when newly exposed by mining), and that it weakens at high temperatures. So solvent saturated with oxygen will be warmed by this reaction, and the hotter it gets the faster this liquid-phase combustion will go. And the huge hydrostatic pressure at the bottom of a solvent-flooded mine will raise the boiling-point of the solvent to above 300 °C, weakening the coal drastically. When the solvent finally boils, the discharge of displaced liquid from the outlet shaft will remove the hydrostatic pressure, and the superheated liquid underground will boil in a vast spectacular explosion on the principle of the natural geyser. The coal will be stripped from its seam in this furious turbulence, and a huge eruption of boiling fluid and coal fragments will vomit out of the shaft. Cold solvent will then be poured back into it, and the cycle will repeat. Not only will the National Coal Board get its coal painlessly; it may even gain for the dour pit-country some of the tourist traffic of Yellowstone Park!

(*New Scientist*, 2 November 1978)

From Daedalus's notebook

The density of coal is on average about 1400 kg m^{-3}, less than that of perchloroethylene dry-cleaning fluid (1620 kg m^{-3}). So coal will float in dry-cleaning fluid! Hence:

Solvent-mining. You just sink two shafts, one at the lowest point of the seam of interest, the other at the highest. Connect them by a single tunnel driven through the seam. Then pump dense solvent down the deeper shaft, and collect the dislodged coal that comes floating up the other one.

Oxidative solvent-mining, for deeper mines. Consider a deep mine-shaft, say 1000 m deep, flooded with perchloroethylene (C_2Cl_4). The pressure at the bottom will be $p = \rho gh = 1620 \times 9.81 \times 1000 = 1.6 \times 10^7$ N m^{-2} or about 160 atmospheres. The solubility of oxygen in C_2Cl_4 at 20 °C and 1 atm is 0.19 vol/vol, i.e. 1.67×10^{-4} g/g. At 160 atm it will be about 160 times as great, i.e. 0.03 g/g.

How hot will this solvent get if all its oxygen reacts in liquid-phase combustion with the surrounding coal? The oxidation of coal is roughly:

$$C_{7.5}H_5O_{0.3}(0.1 \text{ kg}) + 8.6O_2(0.275 \text{ kg}) \rightarrow$$
$$7.5CO_2(0.33 \text{ kg}) + 2.5H_2O(0.045 \text{ kg});$$
$$\Delta H = -3.5 \text{ MJ}$$

0.275 kg of oxygen will be contained in $m = 0.275/0.03 = 9.2$ kg of C_2Cl_4 at 160 atm. So taking the average thermal capacity of C_2Cl_4 over the temperature rise as $C = 1000$ J kg^{-1} deg C^{-1} and ignoring the 4% or so of CO_2 and H_2O in the solvent, the temperature rise as this oxygen reacts with coal will be $\Delta T = -\Delta H/mC = 3.5 \times 10^6/(9.2 \times 1000) = 380$ deg C. So starting from 20 °C it will end up at 400 °C!

This is very hopeful. The boiling point of C_2Cl_4 at 1 atm is 121 °C, and its critical temperature is only 340 °C. So a single pass of oxygenated C_2Cl_4 will set the mine up as a potential geyser, with superheated supercritical solvent under pressure at the bottom of the shaft. Supercritical solvents are really powerful disintegrants and reaction media for coal (some of the modern oil-from-coal hydrogenation processes use them). To blow the geyser on demand, pump extra oxygen down a pipe to the outlet shaft. Enhanced local liquid-phase combustion will raise the temperature until the bottom pressure exceeds 160 atm, when the coal-bearing contents of the shaft will be forcefully ejected. All the liquid in the mine will surge along to replace it, stripping more coal from the seam; top up the inlet shaft with recycled liquid and let the cycle repeat.

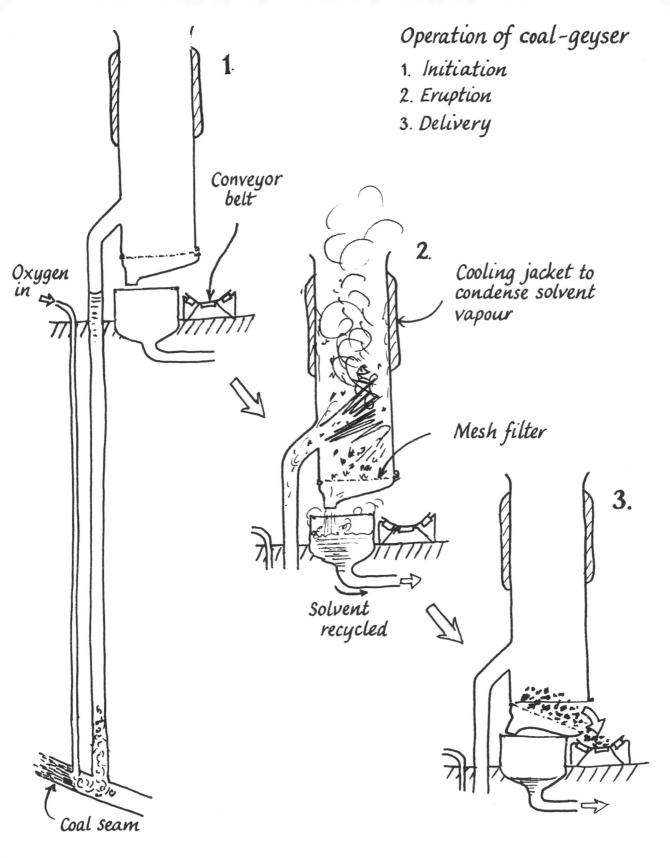

Operation of coal-geyser

1. Initiation
2. Eruption
3. Delivery

1.

Conveyor belt

Oxygen in

Coal seam

2.

Cooling jacket to condense solvent vapour

Mesh filter

Solvent recycled

3.

Drilling with vapour

The drilling of holes is one of the least satisfactory of engineering operations. Deep holes, small holes, and non-circular holes all pose awkward problems for the conventional rotary drill. Seeking a better method, Daedalus recalled that the impact of a shell on armour-plate often detaches a 'scab' from the inner surface of the plate. The reason for this can be seen by making a row of pennies, and rapping another penny smartly into the line from one end. The penny at the far end is cleanly detached; a sequence of collisions has run down the row. If two pennies are brought up to hit the line together, two pennies move off at the far end; a double set of collisions has run down the row. Since armour-plate is made up of atoms, the impact of a shell must transmit millions of atomic collisions through the armour to knock off that number of atoms from the far side.

Now, says Daedalus, imagine the process scaled down to the single-atom level. One atom of iron hitting an iron plate should condense on it, and a chain of atomic collisions should run right through the plate and evaporate the corresponding atom on the far side. Atoms are frictionless, so no energy can be dissipated no matter how many collisions are involved; furthermore a chain of collisions is self-focusing so the energy cannot spread out. So Daedalus's atomic drill will direct a tight beam of metal vapour at a metal surface. The speed of iron atoms as vapour is about 800 metres per second, well in excess of artillery-shell speeds; but a beam of positive ions would probably be even better. They could be steered, aimed and focused by standard mass-spectrometric methods, and if the workpiece were charged as a cathode they could be made to hit it at high speed and in an accurately parallel beam defined by the field lines. A column of condensate the width of the beam would build up on the surface as the atoms were brought to rest; and a hole would tunnel into the far side of the workpiece as the corresponding atoms on that side were evaporated. The atomic-drill operator will see a core of metal being apparently extruded into the beam until it falls out of the hole. The beam of evaporated atoms from the back of the workpiece will, of course, have the same shape, direction, and energy as the one that went into the front; so it can be used to drill another hole in another workpiece placed behind the first. In fact any number of holes can be drilled from one original beam, through workpieces stacked one behind another. The cross-section of the hole will simply be that given to the beam. So the atomic drill will be most useful for drilling non-circular holes, especially long fine holes that could be made in no other way. Even holes a few atomic diameters across would be no problem.

(*New Scientist,* 18 September 1969)

From Daedalus's notebook

The implications of a train of atomic collisions transmitted through a solid have been studied by R. H. Silsbee (*Journal of Applied Physics*, Vol. 28(11), 1957, p. 1246). He reaches the remarkable conclusion that a chain of spheres of radius r and separated by a distance d will 'focus' a chain of collisions if $d < 2r$ (a condition always met in solid lattices). A small misalignment in the original direction of impact is reduced by successive collisions until, far down the chain, they are all exactly in line. This seems borne out by experiments with pennies, too. So thermal displacements, small imperfections, etc., which are inevitable in any solid, should not prevent the long-distance transmission of collisions. Even so, it would seem sensible on general grounds to work the process at the lowest possible temperature. Since it will have to be done in vacuum anyway, to generate and steer the atom/ion beams, this should be no great problem.

A focused chain of collisions

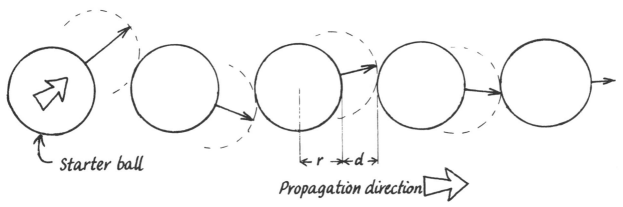

Starter ball

← r → ← d →

Propagation direction ⟹

Successive collisions 'focus' towards perfect in-line propagation if $d < 2r$.

Atomic - vapour drill

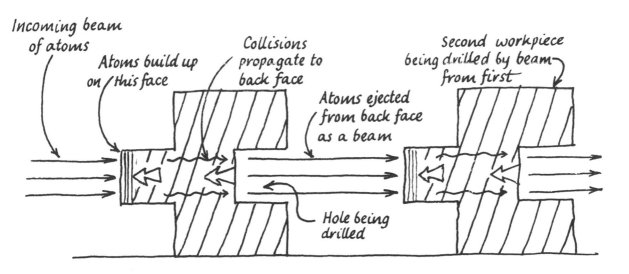

Incoming beam
of atoms

Atoms build up
on this face

Collisions
propagate to
back face

Atoms ejected
from back face
as a beam

Second workpiece
being drilled by beam
from first

Hole being
drilled

Distributed distillation

Drinkable piped water is a scarce luxury in many parts of the world, and is increasingly expensive even in Britain. Daedalus recalls a suggestion to save it by duplicating the pipe network with a parallel system carrying seawater for lavatory flushing and other non-critical uses. He now presents a much better idea, based on those gabardine raincoats which don't let liquid water through, though they are permeable to the vapour. Suppose, says Daedalus, you divided a water-pipe into two 'storeys' by a horizontal lengthways partition; and then passed seawater through the upper half and freshwater through the lower half. The water-repellent fabric would keep the two safely separate. But if the upper seawater were only 0.4 deg C warmer than the lower freshwater, its vapour-pressure would exceed that of the freshwater, and pure water-vapour would then distil from the seawater through the gabardine to augment the freshwater supply. And a half-buried pipe, its upper surface warmed by the sunlit atmosphere, will easily sustain more than $0.4\,^{\circ}\text{C}$ of temperature difference from top to bottom.

The many thousands of kilometres of pipe in a national water system would then constitute a vast, fuel-less, distributed distillation system, accepting seawater at its inlets and delivering freshwater to its customers. (They would also get a simultaneous compulsory supply of extra salty seawater-concentrate from the upper half of the piping system, for flushing their lavatories.) Daedalus is also devising a seagoing version of the same technology: his 'marine dewpond'. This is essentially a gabardine jollyboat to be towed behind the mother vessel at night. Small objects exposed to the radiation-sink of the night sky rapidly cool down, whereas the massive sea maintains its daytime temperature. So water-vapour will distil through the fabric and accumulate as freshwater in the bilges of the vessel, to be pumped out at dawn. If too efficient, however, this ingenious invention will inevitably fill up and sink.

(*New Scientist*, 9 August 1979)

From Daedalus's notebook

Seawater contains 3.6% of dissolved solids, mainly salt. Assuming for simplicity that it's all salt, then 1 kg of seawater contains 964 g (53.5 moles) of water and 36 g (0.615 moles) of salt. The salt is fully ionized to Na^+ + Cl^-, so effectively there are 1.23 moles of ions, making the molar fraction of water present $x_W = 53.5/(53.5 + 1.23) = 0.978$. So by Raoult's Law the vapour-pressure of seawater should be 0.978 (97.8%) of that of freshwater at the same temperature.

The variation of vapour-pressure of a liquid is given by the Clausius–Clapeyron Equation:

$$d\ln p/dT = \Delta H/RT^2$$

where ΔH is the latent heat of evaporation of the liquid, R is the gas constant, and T the absolute temperature. For water at $20\,^{\circ}\text{C}$, $\Delta H = 44\,200$ J mol^{-1} and $T = 293$ K so that

$$d\ln p/dT \simeq \delta\ln p/\delta T$$
$$= 44\,200/(8.314 \times 293^2)$$
$$= 0.0619\ \text{K}^{-1}$$
$$\text{i.e. } \delta T = \delta\ln p/0.0619\ \text{K}$$

To bring the vapour-pressure of seawater up to that of freshwater we must raise its temperature by an amount δT sufficient to raise its vapour-pressure by a factor $p_2/p_1 = 1/0.978 = 1.0225$. So $\delta\ln p$, which equals $\ln p_2 - \ln p_1$, i.e. $\ln(p_2/p_1)$, is $\ln 1.0225 \simeq 0.0225$. Hence:

$$\delta T = 0.0225/0.0619$$
$$= 0.36\ \text{deg C}$$

Air-ground temperature differences are many times greater than this, so the system will clearly work. Indeed, with only $2\,^{\circ}\text{C}$ of temperature-difference it should be possible to evaporate 80% of the water out of seawater and leave a residual salty concentrate in the upper pipe containing 16% dissolved solids. Hence with luck the system may deliver 4–5 gallons of fresh water for every gallon of concentrate the consumer must accept for lavatory flushing, etc.

A snag: at night, especially clear cold nights, the temperature-difference will be reversed. The top of the pipe will get colder than the bottom, and the fresh water will evaporate back to dilute the seawater. I suppose one could turn all the pipes upside-down by some rotary mechanism, so that the freshwater was now at the top and continued to have the lower vapour-pressure, but it seems a cumbersome solution. I shall have to invent some greenhouse-effect covering for the pipes, which accepts high-temperature radiation from daylight to warm the top segment, but which doesn't emit low-temperature radiation at night and so prevents nocturnal cooling.

DISTRIBUTED DISTILLATION SYSTEM

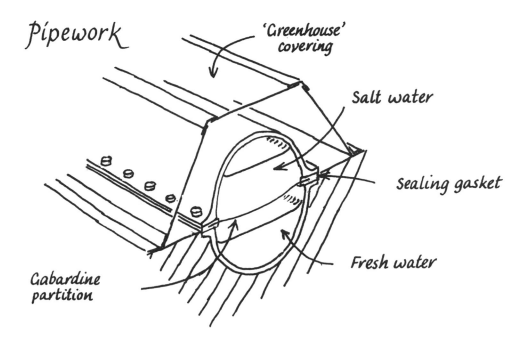

Pipework

'Greenhouse' covering

Salt water

Sealing gasket

Fresh water

Gabardine partition

Domestic interlock to ensure proportional consumption of salt water

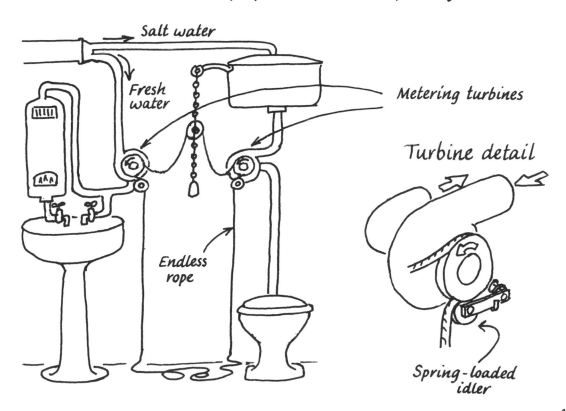

Salt water

Fresh water

Metering turbines

Turbine detail

Endless rope

Spring-loaded idler

The iceberg cometh

A scheme worthy of Daedalus himself has recently been put forward by American glaciologists and geophysicists. They propose to tow icebergs from Antarctica to Australia, where they will be melted for irrigation purposes. Tabular bergs $6\frac{1}{4}$ miles long, $1\frac{1}{4}$ miles across and 800 feet deep are suggested; they would have to be towed by vastly powerful, probably nuclear, tugs. Daedalus reckons that the scheme as it stands is rather technomaniacal; he proposes to make it entirely feasible by arranging for the icebergs to propel themselves, for free. In principle, the temperature-difference between the ice and the surrounding ocean could power engines. For such a tiny temperature-difference the maximum possible efficiency would be under 4%; even so a ton of ice melting every second could give 12 MW of useful power. Since a giant iceberg might require 500 MW to propel it, it would have to be melted at some 40 tons a second to do the job; but this corresponds to a shrinkage of only 2 micrometres per second over all the submerged surface, about the natural melt-rate.

At first Daedalus thought of developing the power by studding the berg with thermocouples to drive electric motors, but he soon abandoned this cumbersome proposal for a more elegant scheme. Freshwater from the melting berg is lighter than the salty ocean water and wells up round the berg. By shaping the stern into a channelled wedge-shape sloping upwards, local melt-water would be induced to rise up the slope and flow out the back, creating thrust by its backwash. Once the berg began to move, all its melt-water would be entrained towards the stern, augmenting the effect, and lengthening and deepening the thrust-creating channel. DREADCO hydrodynamicists are therefore studying the theoretical attrition-regimes of icebergs melt-propelled in warm salty water, to discover the optimum initial stern-contour, whether mid-course reshaping will be needed, and whether a rudder or some differential melting system will provide the better steering. This appealing scheme opens up a whole new field of marine transport. A big berg lasting only a few years but free, expendable, unsinkable, consuming no fuel, and capable of carrying 10 million tons of cargo, might well be competitive with expensive freighters of 10 times the life but only a thousandth of the capacity. Self-propelled icebergs could come to dominate the bulk carrier trade, though with considerable impact on the marine collision problem, and much heart-searching at Lloyd's.

(*New Scientist*, 12 July 1973)

From Daedalus's notebook

Power available from melting ice. To generate useful power P watts in a thermal engine of efficiency η, we need a heat-flow $Q = P/\eta$ W. If this is to come from melting ice, its melt-rate must be $\dot{m} = Q/\lambda = P/\eta\lambda$ kg s^{-1} where λ is the latent heat of melting of ice. With the ice at 0 °C (273 K) and the surrounding seawater at (say) 10 °C (283 K), the maximum possible efficiency of the engine is $\eta = (T_{\text{'boiler'}} - T_{\text{'condenser'}})/T_{\text{'boiler'}} = 10/283 = 0.035$ or 3.5%. Accordingly, since $\lambda = 3.3 \times 10^5$ J kg^{-1} and we want (say) $P = 500$ MW, the melt-rate has to be:

$$\begin{aligned} \dot{m} &= P/\eta\lambda \\ &= 5 \times 10^8/(0.035 \times 3.3 \times 10^5) \\ &= 43\,000 \text{ kg s}^{-1} \\ &= 43 \text{ ton s}^{-1} \end{aligned}$$

A berg 10 km × 2.5 km × 250 m has a submerged area of about $A = 5500 = 250$ m^2 (*sides*) $+ 10^4 \times 2500$ m^2 (*bottom*) $= 2.6 \times 10^7$ m^2. So since ice has a density $\rho = 920$ kg m^{-3}, this melt-rate implies a linear shrinkage of $\dot{x} = \dot{m}/\rho A = 43\,000/(920 \times 2.6 \times 10^7) = 1.8\,\mu$m s^{-1}; quite modest really.

The corresponding heat-flux per unit area at the surface of the berg will be:

$$\begin{aligned} Q' &= Q/A \\ &= P/\eta A \\ &= 5 \times 10^8/(0.035 \times 2.6 \times 10^7) \\ &= 550 \text{ W m}^{-2} \end{aligned}$$

Heat-transfer rates in water/water heat exchangers are several times this even for temperature differences much less than 10 deg C. So in fact the natural melt-rate should be much greater than 43 ton s^{-1}, enabling the required 500 MW of useful power to be generated even if the efficiency of melt-propulsion falls below 3.5% in practice, as it certainly will. A berg of these dimensions will weigh about $m = 6 \times 10^{12}$ kg, so my calculated melt-rate gives it a life of about $t = m/\dot{m} = 1.4 \times 10^8$ s = 5 years or so. The American workers seem to think it will last at least 1 year, which is hopeful.

Shaping a berg for self-propulsion. There are two possible ways:

(a) Blast an angle at the rear end of the berg, under the water, and wait for melt-water thrust to develop.

(b) Put a large weight at the bows. This would angle the underside of the whole berg and initiate preferential outflow of melt-water at the stern. Once again, the berg would be self-optimizing once it was moving. Furthermore, the weight could subsequently be shifted around on the berg for steering. Unfortunately, many millions of tons would be needed, so response would be rather slow.

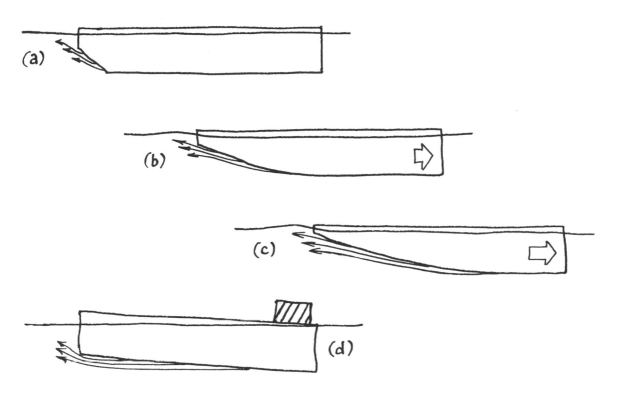

(a), (b), (c) Successive development of underwater contour and meltwater thrust in an iceberg given an initial angle at the stern.
(d) Use of a weight at the bows to develop meltwater thrust over the entire bottom surface as an inclined plane

Daedalus comments
This scheme was triggered by a report by N. Hawkes in *The Observer* for 3 June 1973 (p. 10). He outlined a paper by W. F. Weeks and W. J. Campbell in *Science and Public Affairs* on towing icebergs of the size I quoted from Antarctica to Australia or South America for irrigation purposes. There have since been several other proposals of this nature. In 1980, *Annals of Glaciology* (Vol. 1) published the proceedings of a serious Cambridge conference on the use of icebergs and its scientific and technical feasibility. It even included papers on melt-attrition regimes in moving icebergs — though the authors (H. E. Huppert and J. F. Nye) were concerned to find out whether a towed iceberg would capsize, rather than whether it could propel itself.

But I cannot leave this topic without paying tribute to its true founder, that irrepressible inventor and scourge of the British Patent Office, Arthur Paul Pedrick. Among the hundreds of eccentric inventions which he forced on the embarrassed Patent Office during the 1960s and 70s was BP 1 047 736, granted on 15 October 1965. It outlined a scheme to irrigate Australia by rolling enormous snowballs from the Antarctic down a vast system of intercontinental pipes set up for the purpose. I can only stand in silent awe.

Hollow molecules

There is a curious discontinuity between the density of gases (around $0.001 \, \mathrm{g \, cm^{-3}}$) and that of liquids and solids (from 0.5 to $25 \, \mathrm{g \, cm^{-3}}$). Daedalus has been contemplating ways of bridging this gap, and has conceived the hollow molecule. This would be a closed spherical shell of a sheet-polymer like graphite, whose basic molecule is a flat sheet of carbon atoms bonded hexagonally rather like chicken-wire. He proposes to modify the high-temperature synthesis of graphite by introducing suitable ill-fitting foreign atoms or molecular units into the sheets to warp them (rather like 'doping' semiconductor crystals to introduce discontinuities). The curvature thus produced in the sheet will be transmitted to its growing edges so that it will ultimately close on itself. The radii of the molecules thus produced would be controlled by the level of impurities included in them. Daedalus calculates that a substance made of hollow molecules 0.05 micrometres across would have a bulk density of about $0.04 \, \mathrm{g \, cm^{-3}}$, about half-way between the densities of liquids and gases, and should constitute a vague fifth state of matter. These enormous molecules (molecular weights up to 100 million!) could hardly evaporate, but would interact so weakly at their few points of contact as not to be solid or even liquid. They should behave as tenuous fluids, retainable in open vessels but without any definite surface, and if heated would expand steadily, without boiling, into a gas-like state.

Such fascinating materials would find a host of uses, in novel barometers and shock-absorbers and fluidization-systems and so on; they might even be ideal as low-drag lubricants, where the rolling contact of the molecules would lower the friction even further in ball-race fashion. Daedalus was worried that they might deform under load until he realized that if synthesized in a normal atmosphere they would be full of gas and resilient like little footballs. So he is seeking ways of incorporating 'windows' in their structure so that they can absorb or exchange internal molecules, thus acting as super molecular-sieves capable of entrapping hundreds of times their own weight of such small molecules as can enter the windows.

(*New Scientist*, 3 November 1966)

From Daedalus's notebook

Can a hollow molecule be stable? It might easily deform and collapse to fill the aching void within it:

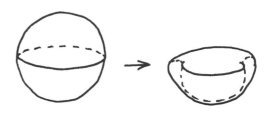

For the molecule to be safe against this mode of inversion, the energy supplied by the attraction of the two sides as they come together must be less than the energy needed to cause the buckling round the equator. So let's work out what these energies are.

(a) The force needed to hold a chemical bond out of its equilibrium angle is $k(r\theta)$ where r is the bond length, θ is the angle of deformation and k is the 'stiffness' or bending force-constant of the bond. So the energy needed to bend a bond from its equilibrium position out to an angle of deformation θ is $\frac{1}{2}k(r\theta)^2$. To invert a hollow molecule of radius R, we need to bend all the bonds round the edge through 180° (π radians). With a circumference of $2\pi R$ and a distance between atoms of about r (the characteristic bond-length of the atomic sheet-network), there will be some $2\pi R/r$ bonds to be bent, each through π radians; requiring a total energy input of $E_b = (2\pi R/r) \times (\frac{1}{2}kr^2\pi^2) = \pi^3 krR$.

(b) The energy made available when the molecule inverts is the energy released by the merging of the two surfaces, with a loss of $4\pi R^2$ of free surface. If the surface-energy is γ, this energy is $E_s = 4\pi R^2\gamma$. As you'd expect, the surface energy E_s available to cause collapse goes up as R^2, whereas the buckling energy E_b opposing it goes up only as R. So as the molecule gets bigger, E_s must ultimately overtake E_b and collapse becomes inevitable. When $E_s = E_b$ the molecule is just stable, and has the largest possible radius R_{max}, obtained by equating our expressions for these energies:

$$4\pi R_{max}^2 \gamma = \pi^3 krR_{max} \qquad (1)$$

$$R_{max} = \pi^2 kr/4\gamma$$

(This is rather inexact. Instead of one bond bending through 180° at the equator to make a sharp fold, several successive bonds will each bend through some lesser angle. This will reduce the total energy of bending. On the other hand, the resulting rounded lip will enclose some free surface, so that not all the available surface energy will be released. These two effects will oppose each other, so the overall error may not be too great.)

Now let's put into (1) some values appropriate to a graphite sheet-molecule. $r = 1.4 \times 10^{-10}$ m, $k = 20$ N m^{-1} (both of these come from the equivalent values for the benzene molecule, which in a sense is a single hexagon from the graphite lattice), and $\gamma = 0.3$ J m^{-1} (a rough guess at the surface energy of graphite along the sheets). We find:

$$R_{max} = \pi^2 \times 20 \times 1.4 \times 10^{-10}/(4 \times 0.3)$$

$$= 2.34 \times 10^{-8}\text{ m}$$

This is quite a size for a hollow molecule: about 0.05 μm across, some 330 atomic diameters! The molecule will have about 260 000 carbon atoms, molecular weight $12 \times 260\,000 = 3.1 \times 10^6$. Assuming each molecule occupies a cube 5×10^{-8} m on a side, the bulk density of the resulting substance will be (mass)/(volume) = $(3.1 \times 10^6 \times 1.67 \times 10^{-27})/(5 \times 10^{-8})^3 = 40$ kg m^{-3}. This is pretty light — about 0.04 times the density of water. Bigger hollow molecules should also be possible, with densities down to 5 kg m^{-3} and molecular weights in the hundreds of millions, but they will have to be stabilized against collapse by internal gas pressure.

Theory of polyhedral molecules. Euler's Law states that for any polyhedron, (no. of corners) + (no. of faces) − (no. of edges) = 2. This prevents any polyhedron being made up entirely of hexagons, a network of which has C + F − E = 0. In that wonderful book *Growth and Form* (Cambridge University Press, pp. 708 and 738), W. D'Arcy Thompson discusses this problem in connnection with radiolaria, those microscopic sea-creatures whose silica skeletons are frequently made up of hexagonal meshes. Even the

beautifully symmetrical *Aulonia hexagona* (which is almost a perfect 100 000-fold scale enlargement of a 1200-atom hollow graphite molecule) has some non-hexagonal faces:

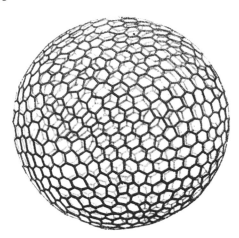

Aulonia hexagona Hkl, *c.* × 200. (From *Growth and Form* by W. D'Arcy Thompson, Courtesy of Cambridge University Press)

Thompson shows that a hexagonal grid, no matter how big, can be closed into a polyhedron by the inclusion of exactly 12 pentagons in its network. So if I warp my graphite sheets by 'doping' them with a pentagon-inducing impurity, there must be exactly 12 impurity moieties for every hollow-molecule's-worth of carbon atoms. This is a very neat way of controlling the size of the resulting hollow molecules. Thus to obtain maximum-radius molecules of 260 000 carbon atoms each, the doping level must be 12 in 260 000 or 46 p.p.m. on a molar basis. This is quite high compared to semiconductor doping levels.

Daedalus comments

Hollow-molecule chemistry has not come very far since I published this idea. The current record in hydrocarbon chemistry is a mere pentagonal dodecahedron, described by L. A. Paquette and co-workers in *Science* (Vol. 211, 1981, p. 575). Keep at it, fellows!

Money peculiar

The old-fashioned 'gold standard', which limited the pound notes in circulation to the value of the government's hoard of gold, had one striking advantage. It curbed the government's ability to create new money at whim. The advent of the digital computer suggests to Daedalus a new way of controlling inflation: a digital standard. Generalizing from the fact that every pound note has a serial number, Daedalus advocates that all the pounds in existence should be recorded by serial number in a central computer. Each bank-account or store of non-physical money would have to specify not merely the number of pounds held, but their actual serial numbers. Similarly, all accounting would have to state not merely the pounds that have been spent, transferred or acquired, but also which ones. This vast increase in accounting complexity is, however, the sort of numerical routine that modern financial computing systems should be able to take in their stride. Not only will knowledge of monetary circulation patterns be greatly sharpened while forgers face terrible new hazards; inflation will be under rigorous control. For to create new money the government or the Bank of England will need new serial numbers. But computing systems set a definite upper limit to the number of digits in an acceptable number. By building a fixed digital length into the software from the outset, monetary inflation will simply not be possible. The awful upheaval of rewriting all the software to find room for more serial numbers would be too frightful to contemplate.

But, muses Daedalus, these days inflation may actually be necessary and desirable. It keeps people happy with the illusion of rising prosperity and material standards without the pollution and resource-depletion of the real thing. The snag of the present system is that the government runs inflation by printing or creating all the new money itself, and then uses it to buy things. This gives it an unfair advantage compared to everybody else. Daedalus's digital system, if set up with vast excessive space for serial numbers, would do things very differently. It is modelled on the 'continuous creation' theory of the Universe, in which new hydrogen atoms appear randomly throughout space. By analogy, new money would just pop up everywhere all the time: a random-number routine would insert new serial numbers among the old ones in all bank accounts. As in some developments of the cosmological theory, the density of new creation will depend on the amount already present: so the larger the bank-account, the more copiously will new money appear in it. Hence all monetary stores would gain the same annual percentage increase; an automatic 'indexing' system that would leave nobody out. Even the piggy-bank-and-mattress brigade could take their hoard along to the bank at intervals for the serial numbers to be noted and the appropriate new notes issued. But Daedalus sees an intriguing further implication. At present, all pounds are in principle identical and interchangeable, like fundamental particles under Bose–Einstein statistics. Once individualized by serial numbers, pounds should henceforth obey Fermi–Dirac statistics, which are far less 'clumpy'. So without the need for punitive wealth taxes, etc., the wider and more socially equitable distribution of wealth would be encouraged.

(*New Scientist*, 12 September 1974)

A recent news item described the indignation of the Australian authorities on finding that a plastic-token coinage minted for the Cocos-Keeling Islands dependency was inflating less than the Australian dollar and was thus appreciating against the mainland currency. Daedalus has a scientific scorn for the wordy battles of economists over the causes of inflation, etc., and sees in this news-item a way of putting economics on a sound experimental basis. Already the Irish and the Scots are complaining of having to use the rapidly-withering English pound-sterling in their transactions; so, says Daedalus, let us develop and formalize this embryonic diversity. Let's have about half-a-dozen independent currencies circulating in Britain — and let each be managed according to the tenets of its associated economic theory! The exponents of a high-wage economy would control a currency in which the wildest pay-claims would be instantly met in newly printed notes. Others of a more puritanical disposition would control a currency whose circulating volume would strictly reflect the volume of available goods, even if that fell to zero because all the workers were on strike for more pay. The government could carefully balance taxation and expenditure in another currency, but indulge in pump-priming deficit financing in yet another. In yet others, interest-rates would be kept low to encourage investment, or high to attract exchange from abroad, and so on. Some of these would be related by fixed exchange-rates, while those of others would be determined by market forces. Any currency or mix of currencies could be used for a transaction by mutual agreement between the parties concerned.

The whole rich confusion could hardly be worse than what we have now, and would be highly instructive to watch. Daedalus expects a sort of war to develop between the currencies. Some would prove unpopular, others would tempt people into their allegiance and then crash spectacularly, still others would prove impossible to exchange abroad. Ultimately only one would survive — or more likely the economists controlling the losing currencies would be forced to save their jobs by adopting the winner's policies. Thus the theoreticians would be bludgeoned into actually learning something, and economics might become a science in time to save us from the final slump.

(*New Scientist*, 30 October 1975)

Floating populations

It is an amusing experiment to take a hydrogen or helium balloon on a string for a ride in the car. When the car brakes suddenly, the passengers are jerked forwards, but the balloon is jerked backwards — it is lighter than the surrounding air and so has an effectively negative mass. An object of exactly the same density as air would not jerk at all. The forces on it would be exactly balanced, giving it effectively a zero mass, and it would be unperturbed by the most violent accelerations. And this is the basis of DREADCO's new crash-protection system for vehicles. A vehicle filled with a medium exactly as dense as the human body (and water is nearly perfect) would protect its occupants completely from the most severe shocks, by giving them an effectively zero mass. To have the vehicle full of water all the time would be inconvenient, but Daedalus points out that the water need only be present during a crash. So DREADCO's prototype safety-car has a foot or so of water under the floorboards. In rapid deceleration it will surge up against the leading wall and cushion the occupants as they are jerked into it.

A similar system for aircraft — whose occupants are at risk from really violent decelerations — is unfortunately penalized by the weight of water which must be carried. Aviation spirit is a poor substitute on several counts, but Daedalus has hopes for the new thixotropic jelly-fuels. Not only are they very hard to ignite accidentally, but they may well have superior cushioning character-istics. Much more immediately practical is DREADCO's novel lift-alarm, which floods the lift instantly if the cable snaps. Daedalus claims that none of these emergency systems will drown their users, for alarmed people always hold their breath instinctively. He is now testing the prin-ciple by dropping volunteer frogmen in large tanks of water from the top of DREADCO's Astrohydraulics Block.

(*New Scientist*, 18 December 1969)

Preliminary DREADCO experiments in dropping volun-teer frogmen in large tanks of water from a height have now been completed. As a result, Daedalus has been forced to modify his theory of last week about water-cushioning. The lungs and airways of the volunteers, being lighter than the surrounding medium, experienced very strong upward forces on impact and gave rise at the very least to severe hiccoughs. He now realizes that for perfect cushioning against acceleration and shock one must be both immersed in and permeated by a medium of one's own density. Such a medium is the high-pressure xenon–oxygen mixture be discussed on 6 July 1967*. It is a breathable gas as dense as the human body and is an ideal filling for vehicles. Not only does it render seats unneces-sary (for passengers float euphorically in it) but the severest jolts or impacts will go unfelt. At first Daedalus planned to pressurize London Underground trains with his mixture, for such trains could be packed to the roof with suspended rush-hour passengers. Both the strain of stand-ing and the ceaseless battering from the lurching vehicles would be cancelled by the unfelt, buoyant gas cocoon.

But Daedalus now realizes that his principle makes all such transport obsolete. The whole tunnel-system could be pressurized, and a 'slurry' of people in breathable flotation-gas could be pumped around it directly, with enormous velocities and accelerations that would go unfelt by the suspended masses. This would be incompar-ably the most capacious urban transport system ever devised: at only 30 m.p.h. one 20-foot diameter tunnel could easily carry a million people an hour! Gravity-dependent clothing may be troublesome for hats will not stay on nor skirts perform their modesty-maintaining functions; and advocates of human dignity may object too. But time and population-pressure are firmly on Daedalus's side!

(*New Scientist*, 25 December 1969)

Daedalus comments

These DREADCO schemes are not as advanced as I thought at the time. It seems that during the later stages of World War II, the Luftwaffe devised a 'g-suit' full of water for pilots of rocket-powered planes; it was intended to stop them blacking-out during sharp turns and other high-acceleration man-oeuvres. In 1958, the Italian physiologists R. Mar-garia, T. Gualtierotti and D. Spinelli subjected pregnant rats to violent decelerations, up to $10\,000g$. The rats themselves died instantly. But the foetuses inside them, having no vulnerable air-spaces to be displaced by the shock and being completely surrounded by amniotic fluid of the same density as themselves, came to no harm and developed normally after being surgically delivered. And also during World War II, that eccentric genius Geoffrey Pyke proposed a people-pipeline to speed up troop movements in Burma. But Pyke's pipeline would have carried the troops in coffin-like cylinders floating in the propellant liquid to be pumped through the pipe; it is only a primitive version of my idea here.

*See p. 128.

Can vegetarianism be taken too far?

'Three acres and a cow', the traditional minimum requirement for a family to subsist in equilibrium with Nature, seems in some ways rather excessive. Milking or eating animals which eat plants is an incredibly wasteful way of exploiting sunlight. But Daedalus recalls that some flatworms, hydras and lichens form 'partnerships' with certain green algae, which are able to live *inside* their host. In return for board and lodging, they photosynthesize glucose and other such products, which their host could not otherwise make himself. He points out how ideal such a partnership would be for man, who gets energy by burning glucose to carbon dioxide and water, while plants take up these products and photosynthesize them back to glucose. Animal nitrogen-wastes are plant-foods too.

At first Daedalus contemplated a special hat with pipework which enabled the wearer to breathe out through a little greenhouse on top, housing a mustard-and-cress flannel to be periodically chewed. Or one might add photosynthetic flora to the normal gut organisms and use fibre optics to pipe light in along the alimentary canal. But for full self-sufficiency, such simplistic schemes fail because they cannot trap enough light. Furthermore they still incur all the wasteful digestive losses of traditional methods of nutrition. So Daedalus began adapting a kidney-machine to pass blood-plasma through a culture of photosynthesizing algae which would remove carbon dioxide and water and replace it with glucose and oxygen. Because the rejection-systems of the two organisms (which defend any creature from being infected by alien organisms) will never make contact with one another, this symbiosis would be entirely benign. Daedalus estimates that a man's entire food and oxygen supply could be met from a mere square metre or so of surface and about a kilogram of algae—a great advance on three acres and a cow. At one stroke the world's food and air-pollution problems would be solved! Simply by being plumbed into a smallish unit exposed to sunlight (DREADCO's prototype looks like a set of sandwich-boards full of greenery) one would become a complete ecosystem in oneself, needing neither to eat nor to breathe, though these luxuries could still be indulged in occasionally. In particular, the user would need to breathe at night. The algae would shut down during the hours of darkness, and while surplus glucose could easily be stored to bridge the nocturnal hiatus, a night's supply of oxygen would fill a large and cumbersome balloon.

While the energetics of this scheme look quite hopeful, its practical inconveniences, such as the need to be permanently plumbed into a set of sandwich-boards, leave Daedalus unsatisfied. He still hankers after a true symbiosis on the lines of the green flatworms which are permeated by green photosynthesizing algae. So DREADCO biochemists are examining these creatures to see how each species prevents its auto-immune response from rejecting the other. One likely answer is that the cellulose walls of the algae are such that only small molecules like glucose and carbon dioxide can get through them; and these, being the common currency of all biochemistry, trigger no immune response. Even so, the DREADCO team is treating its volunteers with radiation and immunosuppressive drugs before attempting to establish photosynthetic cultures in their skin. The idea is that skin, at any rate in its outer layers, is pretty dead and regularly replaced by new tissue growing up from below. So a cutaneous culture of algae won't do any harm, and with any luck its inward encroachment and colonization will be countered by the outward growth of the skin. In any case, the algae won't be able to penetrate far into the skin before the reduced light level chokes them off. Only their rapidly diffusing output of glucose, etc., will penetrate into the blood-capillary layer where the mutually beneficial interchange of glucose and oxygen for carbon dioxide and water takes place.

Thus Daedalus has invented the 'little green man'. Like many of his science-fiction prototypes, he doesn't need food or air, but may find that his alien appearance handicaps him socially. In fact a whole new dimension may be added to colour prejudice. If this turns out to be a serious problem it could be solved by the use of brown algae, whose fucoxanthol pigment can be as effective as green chlorophyll. This should merely give the innocent impression of a healthy sun-tan. But all the body's 1.5 square metres of surface will have to be pressed into service if enough light is to be captured to keep symbiotic man fully self-sufficient. Luckily recent theatrical endeavours have made total nudity almost boringly respectable. Nonetheless, Daedalus hopes to devise 'frosted clothing' to combine modesty and thermal retention with almost complete optical translucence. If properly designed, such clothes should admit photosynthetic light freely while retaining warmth by the greenhouse effect, and may even raise the efficiency of symbiotic man to such levels that he will be able to survive perfectly even in the English climate. Even if not, he will still have such a large measure of internal self-sufficiency that his needs for food and air will be greatly reduced.

(*New Scientist*, 17 and 24 September 1970)

The typical daily energy-requirement of an adult is 3000 kilocalories (say $E = 12$ MJ). If it is all to be supplied photosynthetically over 10 hours of daylight ($t = 36\,000$ s), this implies a power-input $P = E/t = 330$ W. Unfortunately even the finest of the green algae (e.g. *Chlorella*) are only 8% efficient, so to achieve this rate of photosynthesis $P' = 330/0.08 = 4.2$ kW of sunlight must be intercepted. So even in full sunlight of about 1 kW m^{-2}, photosynthetic man will need at least 4 m^2 of surface. This seems to rule out my 'green man' with his 1.5 m^2 of skin, and makes even the algal sandwich-board pretty cumbersome.

However, things may not be that bad. Probably at least three-quarters of the human intake of food-energy is used up merely in keeping warm. Now the 92% of the sunlight not taken up photosynthetically will largely appear as heat in the algal culture, and much of this will be conveyed by heat-transfer straight into the bloodstream of symbiotic man. So with any luck his heat requirement will be completely met by this 'waste heat', leaving only about 3 MJ per day to be provided nutritively by the culture. Hence 1 m^2 of surface should ideally suffice — though in practice it may give only a partial, but still very valuable, degree of self-sufficiency. 1 m^2 of algal culture implies about 1 kg of actual algae. This doesn't seem an excessive load for the 'green man' to distribute about his body, though in a sandwich-board it would be surrounded by heavier culture-medium, plumbing, hardware, etc.

All in all, the external sandwich-board seems more immediately feasible. The algae need only communicate with the bloodstream via a semipermeable membrane allowing exchange of simple molecules like CO_2, glucose, urea, etc. The truly symbiotic 'green man' is more of a challenge. The symbiotic algae-containing turbellarian flatworms (e.g. *Convoluta*) are worth looking into. How do they get their algae? How do the algae divide and what does the flatworm do with the dead ones? What substances are exchanged between the symbionts? Somebody must know . . .

'A special hat, with pipework which enabled the wearer to breathe out through a little greenhouse on top'

Molecular gyrations

Gyroscopes have many important applications in instrumentation, navigation, stabilization and (possibly) anti-gravity machines. It's a pity they need so much precision to make, and require constant power to maintain. Daedalus recalls that in some solids, such as camphor, the molecules are free to spin in the solid lattice: though normally half must spin clockwise and half anticlockwise so any gyro effects must cancel. But whereas a gyroscope may spin at around 10^4 revolutions per minute, molecules at normal temperatures spin at 10^{10}–10^{11} r.p.m., so there is vast gyro-power in these 'rotator solids' if we can tap it. Daedalus calculates that if the molecules in a 10-gram sphere of camphor could all be set spinning the same way, their rotational energy would equal that of the sphere itself turning at 145 000 r.p.m.!

Now you can set a molecule spinning by hitting it with a quantum of microwave radiation — this is the basis of microwave spectroscopy. So Daedalus plans to irradiate camphor with the appropriate frequency of microwave radiation to induce rotation. In the ordinary way, this would set half the molecules spinning one way and half the other; but cunningly Daedalus will use right-handed circularly polarized microwaves. They will induce molecular rotation in one direction only, and so his camphor target will gradually accumulate angular momentum from the microwaves until it is fully saturated. Each little molecule will then be spinning the same way and a totally novel 'gyroscopic solid' will have been created. It will have fascinating properties. Like a gyroscope, the crystals will precess violently when tilted, and will resist any attempt to reorient them. Placed on a table they will waltz endlessly in an eerie gyroscopic manner, and even tilting their bottle will bring a scrabbling, bucking protest from the molecules within. The frictionless molecules will retain their angular momentum indefinitely; even melting will produce a weird gyroscopic liquid very awkward to handle and pour. Gyro-camphor will be ideal for instrumentation, compasses, etc., for it will retain its momentum indefinitely without frictional inaccuracies building up and without drawing any power. And cast in place in ton blocks, it will make a pleasantly perfumed ship's stabilizer. It may also make economic such boons as the gyro-stabilized hat for skaters, slack-wire artists, tottery pensioners, etc.; and stabilized one-legged furniture to save space in the home. And it will make the ideal joke gyroscopic parcel to baffle postmen.

(*New Scientist*, 16 January 1975)

From Daedalus's notebook

There are quite a number of rotator solids whose molecules spin freely in the crystal: camphor, carbon tetrabromide, pentaerythritol, succinonitrile. We want one with a dipole moment so that it can be spun up by electromagnetic radiation, and camphor seems a good one to start with. Unfortunately I can't find its microwave spectrum in the chemical literature; the nearest molecule that has been studied seems to be norbornane-7-one (A. D. Lopata *et al., J. Molecular Structure*, Vol. 26, 1975, p. 85). I'll use it as a model.

Norbornane-7-one   Camphor

A rotating molecule can have only specific quantized levels of energy: $E = BhJ(J + 1)$ in the ideal (e.g. rigid diatomic molecule) case. J is some integral quantum number, and $B = h/8\pi^2 I$ is the rotational constant, with I the molecular moment of inertia. Radiation of a certain frequency v can be absorbed to spin the molecule up to the next energy level: $\Delta E = hv = 2hBJ$ (things get a bit more complicated with big molecules having three different moments of inertia but let's not worry about that). For norbornane-7-one ($C_7H_{10}O = 110$) the moments of inertia I are all around 200 amu $Å^2$. So for camphor ($C_{10}H_{16}O = 152$) I'll guess at I's in the range 300 amu $Å^2$, around 5×10^{-45} kg m^{-2}, which from $B = h/8\pi^2 I$ implies B's around 1.7 GHz. (They're in the 2.3 GHz range for norbornane-7-one.)

Now at room temperature each molecule will have about $\frac{1}{2}kT$ of energy to give to each axis of rotation. So putting $E = BhJ(J + 1) = \frac{1}{2}kT$, then with $T = 300$ K and $B = 1.7$ GHz, we find $J \eqsim 43$: most molecules are in their 43rd allowed rotational state or thereabouts. So to excite them to their 44th we must irradiate with a frequency $v = 2BJ = 2 \times 1.7 \times 43 = 150$ GHz, i.e. microwaves of about $\lambda = c/(150 \times 10^9) = 2$ mm wavelength. So somewhere in the 2 mm waveband camphor at room temperature should have a sharp absorption due to this transition, and if we excite it with right circularly-polarized microwaves of just that frequency, the molecules will repeatedly take up these 'right-handed' quanta of rotation. But as they drop back to their previous level they will emit either right- or left-handed quanta indifferently. So continued irradiation will gradually saturate the sample with right-handed rotational energy, till all the molecules are spinning the same way. (Or I suppose we could start out with camphor at a very low temperature with the molecules hardly spinning at all, and warm it up with appropriate successively higher frequencies of circularly polarized microwaves till it reached room temperature with all molecules spinning right-handedly. Might be quicker.)

How much rotational energy would fully-spinning camphor have? $\frac{1}{2}RT$ joules per mole, of course. So at 300 K a 10-g sample of the stuff will have $E = \frac{1}{2} \times 8.314 \times 300 \times (10/152) = 82$ J of rotational energy. Now if that 10-g sample were a macroscopically spinning sphere, it would have a radius of about 1.33 cm or 0.0133 m (based on density of 1 g cm^{-3}) and a moment of inertia of $I = 0.4\,mr^2 = 0.4 \times 0.01 \times (0.0133)^2 = 7.1 \times 10^{-7}$ kg m^2. And to contain $E = 82$ J, it would have to spin at a rate given by $E = \frac{1}{2}I\omega^2 = 82$ or $\omega = \sqrt{2 \times 82/(7.1 \times 10^{-7})} = 15\,200$ rad s^{-1} or 145 000 r.p.m.! So a solid-state molecular gyroscope will have a lot more power than a conventional spinning one.

The ultimate in heavy breathing

To avoid the dangers of nitrogen narcosis, deep-sea divers breathe various inert-gas/oxygen mixtures, which permit safe working under pressures of many tens of atmospheres. Daedalus points out that some of the inert gases are very dense, and compressing a gas increases its density still further. Indeed, he calculates that at 50 atmospheres the heaviest stable inert gas, xenon, would be as dense as water, so that a man would actually float in it. At this pressure, only 0.5% oxygen would equal the normal atmospheric content, and complete a novel and idyllic environment combining the euphoric effects of skin-diving and free fall, but with the disadvantages of neither. Daedalus proposes to construct large pressure domes containing xenon with airlocks and decompression-chambers for access. Inside, people could at last achieve that childish longing and fly like birds. Water would rise in the medium, and a lake could be formed on the ceiling, into which corks and so on would fall upwards with an inverted splash. The difference in density between the lake and the xenon below it would be so small, however, that the splash and the radiating waves would develop slowly, and travel in a wonderful dream-like slow motion.

Daedalus sees his idea mainly as a recreational and spiritual outlet for mankind. It is also possible that psychoanalysis while adrift in this totally relaxing atmosphere (lit. and fig.) would enable the careworn to unburden their subconscious minds completely, perhaps even revealing the secret sources of the ancient and universal dream of flying? Daedalus, recalling his own past successes in this direction, suspects that it may be an ancestral memory of some mighty classical achievement.

This hypothesis would explain the enigmatic opening to the rare first draft of Coleridge's masterpiece:

> In Xenodu did Kubla Khan
> A stately pressure-dome decree . . .

(*New Scientist*, 6 July 1967)

From Daedalus's notebook

Xenon appears to be the only gas whose density can exceed that of water: its critical density is 1154 kg m^{-3} at $16.6\,°C$ and 58 atmospheres pressure. So at (say) $25\,°C$ and 50 atm it would be a true gas of the same density as water: 1000 kg m^{-3} or so. Could you breathe it? To get the same concentration of oxygen in xenon at 50 atm as in ordinary air, only about 0.5% by volume would be needed, which would hardly affect the bulk properties of the mixture. No problem there. The viscosity of xenon at 1 atm and $20\,°C$ is about $2.3 \times 10^{-5} \text{ N s m}^{-2}$ compared to 1.8×10^{-5} for air, and gas-viscosity is about independent of pressure. So there shouldn't be much problem there, either. The mixture should not be much more difficult to breathe than the high-pressure divers' mixtures, which have much the same viscosity as air. Anyway, if it turns out that respiration loses efficiency in such mixtures, we can always add a bit more oxygen.

One interesting effect will be that voices will drop in tone to a very gruff and rumbly timbre — a sort of 'Goofyspeak', the exact opposite to the 'Donald-Duckspeak' produced by helium–oxygen mixtures. For whereas helium has a speed of sound much higher than air (970 m s^{-1} compared to 331 m s^{-1} at $0\,°C$), xenon has a speed of sound much lower (169 m s^{-1}). So voice-resonances will drop just about an octave.

Daedalus comments

My thinking here is rather out of the mainstream of dense-breathing-medium research. Professor J. A. Kylstra, in *Scientific American* (August 1968, p. 66), gives an intriguing account of experiments in which animals and, in one case, a human volunteer, breathed oxygenated liquids like saline water. The main problem with liquids is their high viscosity (about $10^{-3} \text{ N s m}^{-2}$ for water, some 60 times that of air) and correspondingly low diffusion-rate for dissolved gases, which reduces the efficiency of gas-exchange in breathing and makes it hard to move enough medium in and out of the lungs. As against this, the pressure-requirements are greatly relaxed. About 5 atmospheres suffices to dissolve enough oxygen in saline water for breathing, and some fluorocarbon liquids can hold enough oxygen at only 1 atmosphere pressure.

THE PUBLIC FLOATING BATH

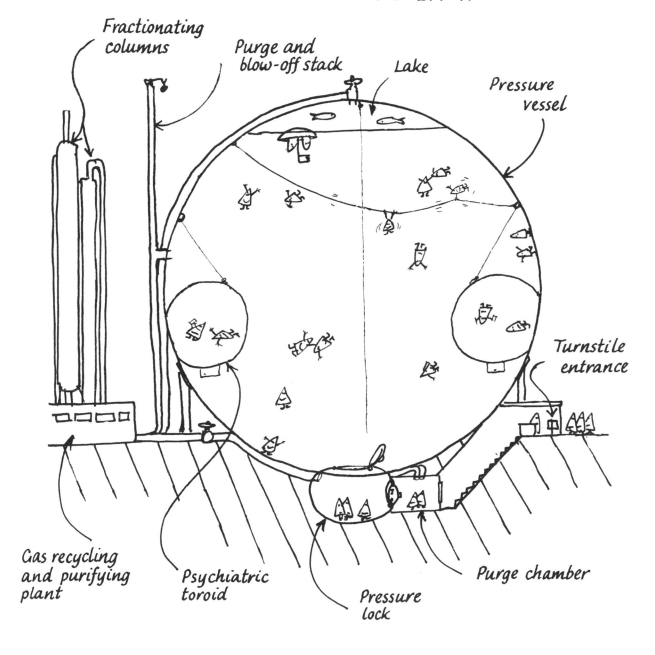

Fractionating columns

Purge and blow-off stack

Lake

Pressure vessel

Turnstile entrance

Gas recycling and purifying plant

Psychiatric toroid

Pressure lock

Purge chamber

Silenced and subverted sound

Daedalus has a new slant on noise-abatement. He points out that any sound is a wave-motion, and can therefore be cancelled by an exactly similar wave motion of opposite phase. So he has been designing apparatus to achieve this benevolent cancellation for such nuisances as jet airliners. His plan is to fit microphones around the jet engines in order to obtain samples of their noise-output. These would feed amplifiers which would invert the phase of the signal and deliver it to loudspeakers, placed so as to produce the exact rarefactions of the air needed to cancel the sonic compressions, and vice versa. The same principle might be adopted to silence motorcycles, pneumatic drills, and so on. But where this cancellation-at-source is undesirable (as in the case of other people's radios, or screaming babies who could well develop lasting complexes as a result of being apparently switched off), Daedalus is developing a portable version to cancel the sound field at the ears of the listener. It would be arranged to be directional so that it cancelled only the undesired noise instead of producing effectively total deafness.

(*New Scientist*, 3 February 1966)

A screaming baby bawls at about the ideal pitch to attract attention. Daedalus wonders if this is deliberate; as a test he plans to flood cradles with harmless gases of various densities and see what happens. If the tot just screeches away instinctively without regard for the results, then helium (which should raise its pitch to ultrasonic inaudibility) or perfluorocyclobutane (which should lower it to a mellow bass) will bring peace to many a tormented family. But Daedalus fears that the crafty brats know what noise annoys a mother most, and will contrive to produce it in any atmosphere by listening to, and adjusting, their own output.

Certainly this process of self-listening and self-correction is important in adults, who can often be reduced to stuttering incoherence by falsifying their audio-feedback with a taped time-delay. There is a rich field here for DREADCO physiologists, who are experimenting with gases and tape-machines, trying to make frogs and parrots stutter and songbirds wander off-key, to discover how closely these creatures listen to themselves. For still more fundamental studies, Daedalus is devising his new stored-program Universal Real-Time Acoustic Filter (URTAF), to enable acoustic feedback to be modified in any way whatsoever.

The first and most basic experiment is to feed back into the ears of the speaker the exact phase-inverse of his speech, and thus to cancel all feedback completely. He will then have no audible clues as to what he is saying. There's an intriguing problem here. Our subjectively perceived voice differs from the objective one recorded by a microphone — many people are very surprised by the sound of their own recorded voice. This is because our own voice is distorted for us by certain resonances in the head. To achieve total subjective-voice cancellation, URTAF will have to incorporate these resonances into the sound it feeds back from its microphone into the subject's earphones. (Incidentally, by feeding the same signal into a loudspeaker, it could also demonstrate to other people what his own voice sounded like to himself.)

Once perfect cancellation has been achieved, and the subject is quite unable to hear what he is saying, the real experiments can begin. For with the cancellation-signal as constant background to remove all natural feedback of the subject's voice, subtly falsified signals can now be added in. They alone will be heard; the subject's audiovocal reflexes will take them as authentic voice-feedback and will be guided by them. Thus a singer, fed with falsely sharpened feedback, will sing flat in an effort to correct this spurious error. Diminished volume should likewise cause her to bellow in misguided self-correction. Rapid changes of fed-back pitch and volume should create all sorts of novel intonations — Daedalus hopes that URTAF, if elaborately enough programmed, could be the ideal linguistic trainer for teaching people to vary their characteristic accent, abandon bad linguistic habits, or polish their powers of mimicry. Indeed, it is hard to set in advance any limits on the powers of voice-modification by false acoustic feedback. A Scotsman might be given a Chinese accent, or a cat made to bark. Even more dramatic, politicians might be forced into the major mind-broadening experience of finding a speech on working-class solidarity emerging relentlessly in favour of feudalism!

(*New Scientist*, 28 May 1970)

When I first thought of sound-cancelling in 1966 I was highly delighted and considered it one of my better notions. So it was rather chastening to discover that I had been anticipated. It turned out that Arthur C. Clarke had invented the idea, and had used it in a story, 'Silence, Please', in his collection *Tales from the White Hart* (Ballantyne Books, 1957). Still, it is something of an honour to be anticipated by Arthur C. Clarke, and honour is doubly satisfied by the fact that this particular scheme has come true.

Several sound-cancelling schemes have since been patented, and indeed they seem to work. The earliest one to my knowledge is described in British Patent 1 304 329 granted to the German aircraft firm Messerschmidt-Bölkow-Blohm GmbH in 1973. It is intended to reduce the noise of rotors or propellers in aircraft or hovercraft by generating 'anti-sound' (*New Scientist,* 31 May 1973, p. 556). To generate enough anti-sound to cancel the very loud output of an aircraft rotor, MBB propose to use a compressed-air-driven siren, or even part of the aircraft fuselage driven in vibration as a loudspeaker.

In 1977 (17 November, p. 427) *New Scientist* reported that departments in three British educational establishments (Southampton and Cambridge Universities, and Chelsea College of Science and Technology) were studying methods of using anti-sound to cancel low-frequency sound in aircraft, fan-driven ventilation ducts, and even free-field sound in the open air. And in 1981 (16 April, p. 165), the magazine reported the first commercial exploitation of the idea — the installation at a British Gas pumping station at Duxford of an anti-sound system with 72 loudspeakers, to deaden the noise of a big gas-compressor!

(a) Original sound

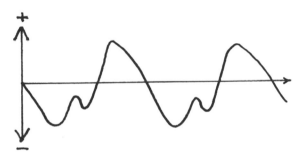

(b) Anti-sound, which at every instant has an equal and and opposite pressure deviation to the original sound. It has the same frequency-spectrum and sounds identical, but if added to the original sound produces:

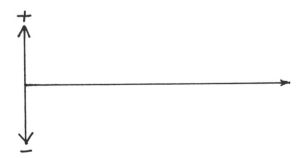

(c) Total cancellation

Tired light

The enigmatic 'redshift' of the light from distant galaxies is usually taken to indicate that they are receding from us at mighty speed. But Daedalus favours the 'tired-light' explanation that light gradually loses energy, and therefore frequency, on its wearisome journey. One snag is that, relativistically, the time-scale collapses totally at the speed of light. Any journey must seem instantaneous to photons, so how can they age or get tired? But, says Daedalus, if light travels not through a vacuum but a refracting medium, it travels more slowly than its classical vacuum 'speed of light'. Its time-scale will still contract, but not to zero. A photon's journey through such a medium will 'subjectively' take time, giving it a chance to age. Now intergalactic light does indeed traverse such a medium — the intergalactic gas. It only has about one hydrogen atom per 10 litres of space, so its refractive index is very close to unity: Daedalus makes it $1 + 2.6 \times 10^{-28}$. Even so, a photon travelling through it for 10^{10} light-years (the distance corresponding to one octave of redshift, i.e. loss of half the photon's energy) would age subjectively by about 2 hours on the journey. So Daedalus believes that photons must have an energy half-life of about 2 hours.

To test this remarkable conclusion he is now devising an experiment. He points out that water has a much higher refractive index than the intergalactic gas. Accordingly, light travels so slowly in water that its time-scale is dramatically extended; it should age by its energy half-life in only 2.7×10^{12} metres, so that it will lose 0.1 parts per million of its energy in a mere 390 km. A frequency-drop of this order is readily detected by laser interferometry, so Daedalus is seeking a site for a straight 400-km optical pipe full of water. The obvious choice is the Nullarbor Plain in Australia, where the railway runs dead straight for 500 km and could easily service an aquatic pipeline alongside. If funds are not forthcoming for this bold attack on the key cosmological puzzle of our time, Daedalus will try to work it into a trans-Australian pipeline scheme.

(*New Scientist*, 9 March 1972)

From Daedalus's notebook

A galaxy d metres away shows a frequency redshift of $-\delta v/v = Hd/v_r$ where H is the Hubble constant and v_r is the velocity of the redshifted radiation. This is conventionally interpreted as a Doppler shift, implying a velocity of recession of the galaxy $V_{apparent} = Hd$. Also, v_r is conventionally taken to be c, the true vacuum speed of light; but since the radiation is traversing the intergalactic medium of refractive index n, we must really put $v_r = c/n$. Taking this law as linear for 'small' cosmic distances, we can write $d = \delta l$, so that:

$$- \delta v/v = nH\delta l/c$$

Putting $L_c = c/nH$, this becomes:

$$- \delta v/v = \delta l/L_c \qquad (1)$$

Now if we interpret this law as showing the 'tiring' of the light, which loses frequency δv while traversing distance δl, we can obtain the total frequency-loss for any finite distance l simply by integrating:

$$v = v_0 \exp(- l/L_c) \qquad (2)$$

Here v_0 is the initial light frequency, and v its frequency after having traversed distance l. When $l = L_c$, $v = v_0/e$. Hence $L_c = c/nH$ is a 'characteristic cosmic distance' over which light decays to l/e of its original frequency. The more familiar '$\frac{1}{2}$-distance' over which light loses half its frequency and experiences a redshift of one octave, is therefore $L_{1/2} = L_c \ln 2 = 0.69c/nH$. Since n is very close to unity, this compares well with the value of $L_{1/2} = 0.6c/H$ given by the conventional Doppler formula with relativistic correction. So far so good.

How long does it take a photon to cover the characteristic distance L_c? In the intergalactic medium of refractive index n, and with corresponding velocity $v_r = c/n$, it will seem to an external observer to take a time $t_c = L_c/v_r$. By substituting c/n for v_r and c/nH for L_c, we get simply $t_c = 1/H$. But to the photon itself, this interval is less. In the photon's frame, the journey-time will be contracted by the relativistic contraction-factor $\sqrt{1 - v_r^2/c^2}$, which with $v_r = c/n$ becomes $\sqrt{1 - 1/n^2}$. So the 'subjective' time it takes for a photon to cover distance L_c and decay in frequency by $1/e$ is:

$$\tau = \sqrt{1 - 1/n^2}/H \qquad (3)$$

My thesis here is that this passage of time *causes* the loss of frequency which we observe as the redshift. More precisely, photons lose frequency exponentially with subjective time rather on the lines of radioactive decay. On this view τ is the 'characteristic decay time' in an equation exactly paralleling (2) above:

$$v = v_0 \exp(- t/\tau)$$

where t is the 'subjective time' which the photon has experienced since being emitted at frequency v_0.

τ must be a fundamental property of photons, and (3) enables us to estimate it from values of H, the Hubble constant, and n, the intergalactic refractive index. So let's try it.

The refractive index of a gas is rather well represented by $n = 1 + kN$, where N is the number of particles per m^3 and k is a constant for the gas. Making this substitution in (3), and recalling that for gases kN is always small compared to unity, we can extract as a close approximation:

$$\tau = \sqrt{2kN}/H$$

For hydrogen at $0\,°C$ and 1 atm, $N = 5.3 \times 10^{25}$ atoms m^{-3} and $n = 1.000\,138$, giving $k = 0.000\,138/(5.3 \times 10^{25}) = 2.6 \times 10^{-30}\,m^3$. N for the intergalactic hydrogen is 100 atoms m^{-3} or so, and the Hubble constant H is about $2 \times 10^{-18}\,s^{-1}$, so we finally obtain:

$$\tau = \sqrt{2kN}/H$$
$$= \sqrt{2 \times 2.6 \times 10^{-30} \times 100}/(2 \times 10^{-18})$$
$$= 11\,400\,s$$

The corresponding photon-frequency half-life will be $\tau_{1/2} \doteq \tau \ln 2 = 7900\,s$; about 2 hours. The dying of the light is really quite rapid!

Can we check any of this experimentally? The foregoing analysis holds equally well for any refracting medium. So for any medium we can define a 'Hubble constant' H_m and a 'characteristic distance' L_m solely from its refractive index n_m and our value for τ:

$$H_m = \sqrt{1 - 1/n_m^2}/\tau$$

$$L_m = c/n_m H_m = c\tau/(n_m\sqrt{1 - 1/n_m^2})$$
$$= c\tau/(\sqrt{n_m^2 - 1})$$

A medium with a high refractive index n_m will have a short characteristic distance L_m; photons will travel so slowly in it that they will spend quite a lot of time, and lose quite a bit of frequency, on even a modest journey. Try it for water whose n_w is 1.33:

$$L_w = c\tau/(\sqrt{n_w^2 - 1})$$
$$= 3 \times 10^8 \times 11\,400/(\sqrt{1.33^2 - 1})$$
$$= 3.9 \times 10^{12}\,m$$

(The corresponding $\frac{1}{2}$-distance, $L_w \ln 2$, is $2.7 \times 10^{12}\,m$.)

So from (1), the loss of frequency over a journey δl in water is given by:

$$- \delta v/v = \delta l/L_w = \delta l/(3.9 \times 10^{12})$$

So a frequency-loss $- \delta v/v$ of 10^{-7} — which could be quite easily detected — should occur over a distance:

$$\delta l = 10^{-7} \times 3.9 \times 10^{12}$$
$$= 3.9 \times 10^5\,m$$
$$= 390\,km!$$

This looks well within practical achievement.

Daedalus comments

Since I made this calculation in 1972, advances in fibre optics and frequency-measurement have made this experiment much more feasible. Detectable redshift should be exhibited by a mere few kilometres of optic fibre.

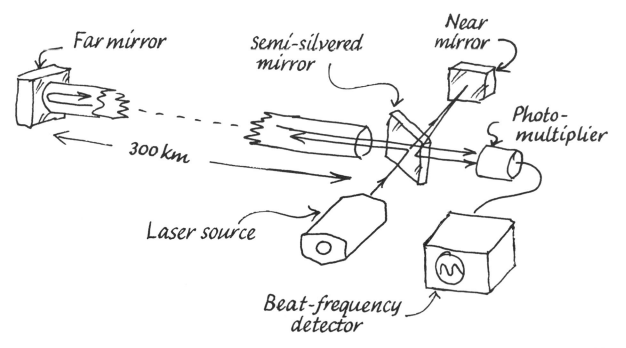

Far mirror   Semi-silvered mirror   Near mirror

300 km

Laser source

Photo-multiplier

Beat-frequency detector

Where there's muck there's history

Daedalus regards pollution as a priceless national heritage. He recalls how the historical expert can isolate, layer by layer, successive inscriptions on a much-used parchment, and is developing similar techniques for the layered filth of northern towns, the mighty heaps of old industrial waste and the fouled bottoms of ancient canals. Every industrial process has its own characteristic effluvium which ultimately settles over the neighbourhood; so detailed chemical analysis of the grime would reveal the fuel and its mode of combustion, the reactants and products, the lubricants used and the fine dust abraded, etc. Every shutdown, every change in the process or its output will be subtly written in the innumerable superposed layers of grime on abandoned factory or grim gothic town-hall. Modern microscopic and microanalytical methods at last enable the industrial archaeologist to read this fascinating story — provided it has not been scrubbed away by some thoughtless clean-environment enthusiast! Daedalus pleads for the preservation of historic dirt in old gasworks and railway tunnels and slag-heaps, in which the fuel and metallurgical history of our country is enshrined. Save our historic grime!

On a smaller scale, the same principle has a fascinating use in studying the history of cooking. Every meal cooked sends forth its aroma to condense on the walls in a delicate, fragrant layer of grease. Over the years, therefore, there should build up on kitchen and dining-room surfaces, in perfect order, the chemical signatures of successive breakfasts, lunches and dinners. They may even be sub-layered in identifiable banquet order from hors-d'œuvre to brandy and cigars! Since the volatiles of a meal define its flavour, they preserve the very essence of the cooking. So delicate chemical treatment of French hotel kitchen ceilings unwashed since the last century should reveal an unbroken succession of menus dating back to the fabled Escoffier himself. Even the secrets of long-lost recipes may be recoverable by careful chemical detective-work.

(*New Scientist*, 7 May 1970)

Daedalus comments

One rather beautiful example of chronologically layered pollution has been known for a long time — the 'Sunday Stones' from the drainage pipes of early coal mines. The waste-water pumped up from the mine always contained white silt and sediment. But while the shift was at work it contained black coal-dust too, from the dust-laden air of the busy workings. The silt which deposited and compacted in the pipes clearly shows the regular pattern of the shift-work and the six-day week in force, as a sequence of successive black lines. The absence of a line at every seventh interval shows that the mine lay undisturbed every Sunday.

My generalization of the principle has yet to be explored fully. But one prophetic detail has already come true. My suggestion that industrial history could be read 'in the fouled bottoms of ancient canals' has been triumphantly taken up by E. Goldberg and his colleagues (*Geochemical Journal*, Vol. 10, 1977, p. 165; *New Scientist*, 31 March 1977, p. 757). These workers decided to study the moat of the Japanese Emperor's Palace in Tokyo. Investigating the palace grounds of a divine mortal is not simple; but after much bureaucratic effort, Goldberg secured a core from the moat in 1975. From its chemical layering he was able to deduce the history of Tokyo's air pollution. The rise in motor traffic density and the later restrictions on exhaust emissions are reflected in changing lead and cadmium levels through the core; a minimum in zinc and copper levels around 1955 may reflect the switch from a war-weakened economy to a peaceful and booming one; very high carbon levels reflect the importance of coal and charcoal as domestic fuels.

So owners of stately homes, don't stir the moat! It may still be compiling its own subtle memoirs of your family's ancestral fortunes . . .

This 'Sunday Stone' is in the Hancock Museum, Newcastle upon Tyne. The six-day week of the mine it came from is evident in the six daily lines of coal-dust, and the Sunday gap showing when the mine was inactive. The second band from the bottom has only five lines — Newcastle United must have been playing at home that Saturday!

HISTORIC MOAT.
NO DABBLING
BY ORDER.

Amplified smell

While telescopes, microphones, etc., have been invented to extend the range of many of our senses, nobody has ever invented a smell amplifier. Daedalus now fills this long-smelt want with the DREADCO 'Meganose'. This simple gadget fastens over the wearer's nose like a large snout. A rapid stream of the surrounding air is blown over a small endlessly-rotating belt cooled to liquid air temperature, so that all odiferous volatiles condense on it. Its rotation takes it through a warming chamber where they are re-evaporated into the wearer's breathing-tube. Because the airstream from which the smell-substances were extracted has a thousand times the volume of the inhaling-air into which they are then released, they reach the nose in a thousand times their natural concentration. This high amplification — which can be made even higher by repeating the process in a second stage — will open up a whole new nasal world. On entering a room we would be immediately aware who was present, who had been there in the past and for how long, by the strength and persistence of their individual odours. A brief sniff would reveal who has left the tap running, or whether the cat really had brought that strange object in. Even the effluvia of overheated wiring, tiny gas-leaks, or a misplaced kipper, might readily be spotted in time to avert disaster.

Tracking down the sources of such smells would present problems, however. Daedalus proposes a two-channel meganose with twin air-intakes to right and left like a bicycle's handlebars. With each intake feeding its respective nostril, the hitherto unknown sensation of 'stereo smell' should result, enabling one to locate instantly the source of the faintest aroma. Clearly this coupling of bloodhound nasal acuity with human intelligence bodes ill for the criminal, who cannot avoid leaving comprehensive aroma-clues all around the scene of his crime. But the scent of the police should also be detectable from some way off . . .

(*New Scientist*, 21 September 1972)

DREADCO's 'Meganose' smell amplifier
(the 'Bloodhound' forensic model)

Close-up attachment for the 'Bloodhound'

From Daedalus's notebook

STEREO MEGANOSE (detail of right-hand channel)

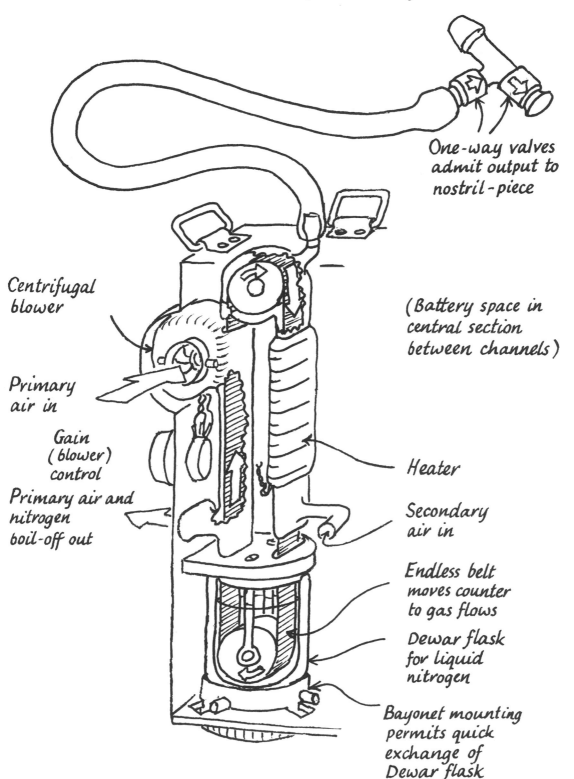

One-way valves admit output to nostril-piece

Centrifugal blower

(Battery space in central section between channels)

Primary air in

Gain (blower) control

Primary air and nitrogen boil-off out

Heater

Secondary air in

Endless belt moves counter to gas flows

Dewar flask for liquid nitrogen

Bayonet mounting permits quick exchange of Dewar flask

Seeing the infrared again

Our colour vision spans the colours of the rainbow from red to violet. But the colours are a mere slot in the full optical spectrum. Most of it, from far infrared to far ultraviolet, we cannot see at all. Daedalus recalls the story of a professor of spectroscopy who lost an eye-lens in an explosion. He was given corrective spectacles with enough UV transparency to let him see some way into this region of the spectrum. Accordingly, he could line up the departmental UV spectrometers by eye, and was much in demand! This story suggests to Daedalus that the retina may well be sensitive to a fair range of UV and IR radiation, if only the eye were transparent to it. In this connection he recalled the aural phenomenon of 'subjective bass'. Cheap gramophones with tiny speakers cannot handle the low frequencies, so their makers equip them with distorting amplifiers which intermodulate pairs of the higher frequencies. The gullible ear is taken in by this spurious imitation. Being a highly non-linear device, it demodulates the intertwined tones and hears an entirely fictitious bass 'difference-frequency' between them. So Daedalus proposes to work the equivalent trick on the eye. Modern laser methods should be able to make a laser beam that swings from yellow to orange (say) and back, at a frequency equal to that of some radiation in the infrared: i.e. to put an IR frequency-modulation on a visible signal. The beam, being entirely in the visible, will penetrate the eye normally. But the retina, just as gullibly non-linear as the ear, will demodulate it and see the implied infrared. A totally new colour, the first major extension of human consciousness, will have been created!

Daedalus cannot imagine, and would not be able to tell us if he could, what his new infrared colours will look like. But his modulated lamps will throw them everywhere, revealing the most banal of objects in a new and enthralling light. Art, display, and ornamentation will all enter a rich new glowing world.

(*New Scientist*, 29 May 1969)

Daedalus comments

This item sparked an amusing correspondence in the columns of *New Scientist* about whether cheap gramophones really did produce 'subjective bass'; if so whether they did it by intermodulation or by introduction of new harmonics, and whether manufacturers deliberately arranged for it or merely took advantage of the inevitable non-linearities of cheap equipment, and the non-linear response of the ear. In hindsight, I should really have proposed putting an amplitude-modulation on my laser beam, not a frequency-modulation. A detector only needs a non-linear amplitude response to demodulate AM; it needs both non-linear amplitude and frequency responses to demodulate FM. However, the ear successfully demodulates even an FM audio signal; it has both kinds of non-linearity, as does the eye. In fact all sense organs have an approximately logarithmic response (Fechner's or Weber's Law), allowing them to register an enormous range of input intensities.

My opening story about the man who could see the UV after losing the lens of his eye was later pleasantly amplified in a letter to *Science* (Vol. 204, 1979, p. 454) from D. Davenport and J. M. Foley of the University of California at Santa Barbara. 'Persons facing lens removal because of cataracts', they wrote, 'frequently view their future with some alarm. To them and in particular to professional colleagues who have this problem, we say, "Cheer up. You'll have advantages you never expected".'

Among the unexpected optical advantages of losing an eye-lens is the enhancement of perceived colour and brightness. Blues in particular are strengthened, and UV also reaches the retina. It is absorbed directly, producing a violet sensation, and in addition causes the retina to fluoresce in the visible, creating a sensation of greenish-blue. With appropriate optical aid, the lensless eye can also produce sharp retinal images of a wide range of magnifications. Professor Davenport (who writes from experience) may be one of the few persons in the world who, when he went to the Tutankhamun Exhibition, took out his contact lenses.

Later I entered this field again, this time with a method of seeing the world in its own IR colours, rather than the 'fake' colour provided by a fixed-frequency modulated lamp. IR-to-visible upconversion had recently been made to work in astronomy (a useful review is given by J. Falk in *Laser Focus*, Vol. 15(10), 1979, p. 72) and seemed ripe for more dramatic applications.

Daedalus has been musing on the fact that cataract surgery often leaves the patient with much enhanced colour vision, extending even into the ultraviolet. The implication is that the retina is sensitive to a very wide range of optical frequencies, if only the eye would let them through. Seeking some way of getting IR radiation through the lens and media of the eye so that the retina could see it, Daedalus recalled the recent successes of 'laser upconversion' in astronomy.

In this technique, IR from an astronomical source passes through a crystal of lithium niobate or silver arsenic sulphide which is simultaneously irradiated by an appropriate visible-light 'pump' laser. Such crystals are optically non-linear. Their refractive index is modulated by one beam, and impresses that modulation on the other beam. So the emerging radiation is the modulation-sum of the two inputs, i.e. visible light carrying an IR modulation. Thus the IR emission of astronomical objects can be conveniently imaged in the visible, often with useful gain into the bargain.

So DREADCO physicists are making a special pair of lithium niobate spectacles. Careful phase-matching and filtering will be needed to keep the pump radiation out of the eye while maintaining the widest possible angle of view. But the effort should be well worth while; for IR entering the spectacles will in effect acquire a visible 'carrier frequency' to carry it through the eye. Once it reaches the retina, the gross non-linearities of retinal detection will demodulate the signal, reinstate the IR, and sense it. Thus a whole new sensory field will be opened up! The everyday world will acquire an incredible richness of unimaginable and highly informative new colours. For while only a few chemical substances show visible colour, almost all of them have strong and highly characteristic IR absorption 'signatures'. So these novel 'Infraspecs' will reveal to the wearer a rich new world of chemical insights coded in amazing and subtle 'infra-colours'. Salt, sugar, chalk and flour will all be dazzlingly different; the strength of gin and vodka will be colourfully betrayed; and adulteration and dilution of all kinds will be embarrassingly apparent to everyone.

(*New Scientist*, 9 April 1981)

From Daedalus's notebook

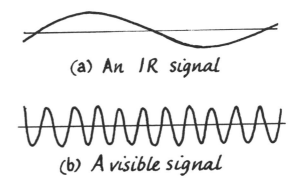

(a) An IR signal

(b) A visible signal

(a) and (b) intermodulate to give:

This is registered on a grossly non-linear detector as:

which contains (as its 'centre of gravity'):

— the original IR signal (a)

The vibro-tram

Most vehicles make constant contact with the ground, and require elaborate suspension mechanisms to operate smoothly. The hovercraft floats freely, but pays for its freedom by having to pump vast amounts of air all the time. So Daedalus has been devising an intermediate sort of vehicle. This machine, inspired by the vibrating conveyor-chute, has in place of wheels large pads or runners under its whole length which are driven in rapid vertical vibration, so that the vehicle makes only intermittent contact with the ground and moves freely in many small jumps. Suitably elastic runners, perhaps made from the rubber used for those 'superballs', would minimize losses so that little power would be needed to maintain vibration.

This ultimate development of the pogo-stick would be able to travel at speeds unlimited by centrifugal wheel stresses, and would be propelled by superimposing a variable horizontal jerk on each downward movement of the runners. Steering could also be achieved this way (though the vehicle might be at its best on rails) and very powerful braking would be available merely by cutting off the vibration. The loading on the ground would be no greater than usual. The road or rail, instead of supporting the load at isolated points in space (the tyre contacts) would do so at isolated points in time. A vibration-frequency of many hundreds of cycles per second would ensure that the individual jumps of the machine would only be a few inches long, and Daedalus hoped at first that such frequencies would be too high to be perceived as such by the passengers. However, he then realized that by the same principle they would be unable to make contact with the floor or seats of the vehicle and would slide helplessly around the vibrating interior. Clearly proper suspension is required to isolate the cabin from the levitating vibration.

(*New Scientist*, 10 November 1966)

From Daedalus's notebook

Suppose we choose 50 Hz mains-frequency for the pilot vibro-vehicle. (Maybe a bit low, but it's readily available and means we can adapt, e.g., a disused tramcar for trials.) Then each jump will take 0.01 s rising and 0.01 s falling. In 0.01 s an object falls a distance

$$h = \tfrac{1}{2}gt^2$$
$$= \tfrac{1}{2} \times 10 \times 0.01^2$$
$$= 5 \times 10^{-4}\,\text{m}$$
$$= 0.5\,\text{mm}$$

So quite short-stroke actuators will suffice for levitation. But the pads will not clear a track irregularity greater than 0.5 mm, so smooth rail would be the preferred track. At each jump the actuators have to impart to the vehicle a vertical velocity of

$$v = \sqrt{2gh}$$
$$= \sqrt{(2 \times 10 \times 5 \times 10^{-4})}$$
$$= 0.1\,\text{m s}^{-1}$$

This is not excessive. But if the vehicle weighs a ton (say 10^3 kg), and this velocity must be given to it $f = 50$ times a second, then the power required is

$$P = \tfrac{1}{2}mv^2 f$$
$$= \tfrac{1}{2} \times 10^3 \times 0.1^2 \times 50$$
$$= 250\,\text{W}$$

Encouragingly economical, especially since we shall recycle most of this power-input through the elastic suspension-mountings. Clearly the power for vibro-levitation is negligible compared to the likely power for propulsion, and will be even lower for higher vibration-frequencies.

Speed. At 30 m.p.h. (say 13 m s^{-1}) and 50 Hz jump-rate, each jump is $13/50 = 0.26$ m $= 26$ cm. Assuming actuators are in contact with the ground 10% of the time, horizontal excursion must be 2.6 cm. Perhaps we should have two sets of actuators, short-stroke for levitation and horizontal long-stroke for propulsion? Or use a single set and vary their inclination?

An adapted tramcar

Two-handle speed and
steering control
varies angle of attack
of vibrators

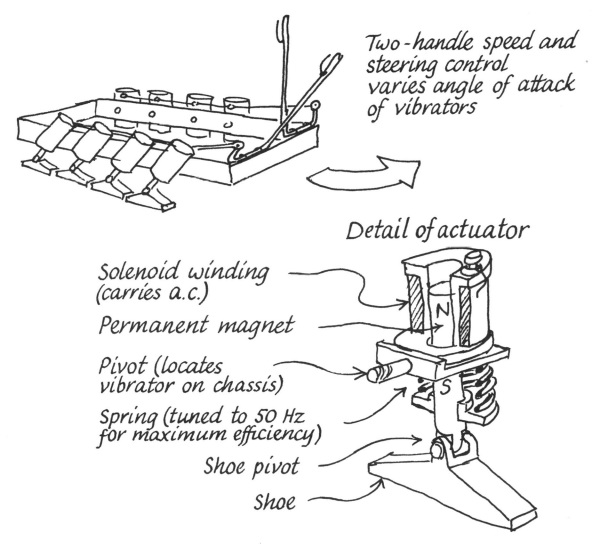

Detail of actuator

Solenoid winding
(carries a.c.)

Permanent magnet

Pivot (locates
vibrator on chassis)

Spring (tuned to 50 Hz
for maximum efficiency)

Shoe pivot

Shoe

N

S

141

Milk of amnesia

Daedalus has been pondering the remarkable assertion that we never really forget anything. Hypnotists can induce detailed recall of long-forgotten childhood events: like some obsessive bureaucracy the brain files copies of *everything*, just in case. So, on the computer analogy, storage-space is steadily filled. As age advances, says Daedalus, storage must begin to encroach into processing space, which is why intelligence and imagination decline. Finally processing has so little elbow-room that malfunctions occur. We become senile, and when the brain is finally full up, we die. Old people who remember long-past events but forget yesterday are clearly delaying their fate by refusing to add to storage.

Now Daedalus recalls that the whole basis of memory, and indeed of all brain-functions, lies in the synapses where the 'output' of one brain-cell connects with the 'input' of another. The signals which cross the synapses are ultimately chemical. An output fibre, receiving a nerve-pulse from its neurone, will release a specific chemical 'neurotransmitter' at its synapse. The receiving neurone can learn to respond to this stimulus by firing off a nerve-pulse of its own, which will spread out along all its output fibres, and may stimulate further neurones to fire in their turn. Tranquillizers are said to work by temporarily blocking the receptor sites for specific neurotransmitters, and some such drugs interfere worryingly with memory. So Daedalus's anti-senile 'milk of amnesia' contains tranquillizer-like molecules so reactive that they not only block, but bind permanently to, the receptors they land on. Memories are thus really wiped out, and a synapse so blocked is freed to 'learn' again via different sites within it, sensitive to a different neurotransmitter. (The brain has lots of them, and it seems not to matter which one a particular synapse elects to use.) Now really important information is distributed in many places and forms around the brain. So milk of amnesia, while it will undoubtedly cause momentary confusion, won't prevent ultimate recall and re-establishment of key knowledge. But the vast amount of trivial bumph that chokes us to death will at last be thrown out! At first Daedalus was worried by the sociological implications of rescuing thousands of old dodderers from the grave and releasing them, spry and imaginative, to compete with the rest of us till kingdom come. But heart-failure, pneumonia, and other syndromes less distressing than senility, will carry them off in the end.

(*New Scientist*, 10 May 1979)

From Daedalus's notebook

Let's assume that the brain, like most computers, stores intelligence (programs) and memory (data) in the same form and distributed throughout the same volume. Then the more space is taken up by data, the less is available for programs and working space. Clearly as life progresses and memories multiply, there must come a time when programs and working space get squeezed. This must be senility.

Consequences of this theory:

(a) Enough novelty will kill anyone. Old people do seem to die more easily if new data are forced on them — death of a spouse, retirement, move into a sunset home, etc. Maybe forgetfulness and protective stupidity do have survival value for the species (no wonder there's such a lot of them about).

(b) There's that intriguing theory that dreaming is the process of coding and repacking the day's experience, in order to fit it into as small a space as possible. People deprived of dreaming get irritable and distracted — is this a pseudosenility? Maybe I could get the psychologists to make comparisons.

(c) Electro-convulsive therapy (ECT) may derive its whole therapeutic benefit from its amnesiac effect. It's just an electric bludgeon and presumably wipes out whole swathes of memory and frees the space for more processing. Now the important memories are duplicated in sites all over the brain (remember those experiments in which trained rats could still run a maze no matter which bit of their brain was removed). But trivial stuff — last year's breakfast menus, etc. — is presumably stored at only one site. So ECT will wipe out a lot of trivia, but now and again will happen to clobber every site of some key memory; the patient will then be aware of a loss. ECT patients do seem to complain of this.

Can we devise a better memory-eraser than ECT? Brain-theory seems to be that memory is encoded in the synapses where the axon of the transmitting nerve-cell touches the body of the receptor nerve-cell. The information stored by the synapse is simply the probability that a pulse from the transmitter will fire the receptor. We can free this probability for re-establishment by blocking the neurotransmitter used to carry the signal, and forcing the synapse to deploy another one. (The brain has dozens of them.) The ideal would be a molecule mimicking the neurotransmitter and blocking its site of action. That way we can knock out just one neurotransmitter at a time, instead of blasting the whole brain with ECT.

A random thought. Tranquillizers are said to work by blocking the sites of selected neurotransmitters. Since they can affect emotion without interfering with reasoning or perception, there's a chance that the brain may code different 'types' of memory — worries, facts, beliefs, etc. — by using different neurotrans-mitters in the networks handling them. If so, and if on this philosophy the brain identifies 'trivial' memories by allocating *them* a specific neurotransmit-ter, then we could wipe out the trivial memories with complete selectivity by subverting that one neurotransmitter. A long shot, but a nice idea.

The surface properties of rubbish

Some interesting principles underlie the recent Russian experiments on welding metals in space. In the Earth's atmosphere, all solid surfaces are contaminated by weakly-bound molecular layers of moisture and adsorbed gas. In the hard vacuum of space, these layers evaporate, and the denuded surfaces can make true molecular contact with each other. Since the molecules of solids attract each other strongly (that's why they *are* solids), the bare surfaces weld together very readily. Indeed, many of the early space-probes failed unexpectedly owing to the seizure of switches, bearings, etc., in the vacuum of space. Daedalus was watching one of those rubbish-compacting dustcarts at work when he was struck by the thought that only poor surface cleanliness prevented the compressed rubbish pressure-welding solid in the cart. He toyed with the idea of an orbiting dustcart to test the notion, but in more down-to-earth fashion set DREADCO's composite-materials group to work on a new vacuum-rubbish treatment. The process is simplicity itself. Incoming rubbish is exposed to high vacuum which flashes off all liquids, leaving every item bone-dry and uncontaminated. The pristine broken glass, anhydrous fag-ends, dehydrated and totally porous potato-peelings are then vacuum-compacted under sufficient pressure to ensure complete collapse, plastic flow, and contact-welding throughout the mass. Junked alarm-clock will weld to dud light-bulb, horsehair cushion will bind indissolubly to battered tin, apple-core and razor-blade will lock in chemical union; and the end-product will be a voidless solid board whose humble origins will be proclaimed by a fascinating surface grain of all these disparate objects locked in surrealistically distorted and interpenetrating patterns.

DREADCO's 'Junkboard' (Regd) should prove very popular. Its all-welded, composite microstructure will undoubtedly give it great strength without brittleness, and its cheapness and rich, detailed surface finish will ensure a wide market. And DREADCO will be doubly happy when, after a long and useful life, it is chucked back into the bin.

(*New Scientist*, 20 November 1969)

Daedalus comments

In the issue of 12 July 1973, *New Scientist* reported (p. 78) that workers at the Warren Spring Laboratory had devised a method of recycling mixed metal and plastics waste to 'a strong, light waterproof material resembling chipboard'. This first step on the road to a true universal Junkboard is intended for packaging, roofing, etc. I wish it well!

Most metals react rapidly with oxygen even at room temperature. We don't usually notice the effect, because the metal surface soon gets coated with a thin film of metal oxide which blocks further reaction. The process is a surface combustion, and gives out the appropriate quantity of heat for the small amount of reaction which manages to occur. But a fine-enough metal powder is essentially all surface. Exposed to oxygen it reacts completely, and its total combustion gives out enough heat to cause fierce flame. Such 'pyrophoric' metal powders, as chemists call them, must be kept under liquids or in inert atmospheres; in the air they ignite immediately. Burning is such a universally easy reaction that Daedalus initially reckoned that everything must be pyrophoric if finely enough divided. But then he recalled that both carbon and petrol burn to carbon dioxide, but finely divided carbon ignites spontaneously (a hazard of charcoal-burning), whereas finely sprayed petrol does not. He now reckons that the shuffling molecules of a liquid can never provide a firm foothold for incoming oxygen molecules, whereas they can bind firmly enough for reaction to the static molecular lattice of a solid with its well-defined and enduring binding-sites. So all combustible solids should form pyrophoric powders.

This suggests a neat solution to the twin problems of waste-disposal and energy-recovery. Rubbish of all sorts — plastic waste, tea-leaves and potato-peelings, beer-cans and old cars: everything, in fact, except glass and china — is theoretically combustible, but hard to burn usefully and completely in practice. So instead of heating the stuff up in incinerators, DREADCO engineers are freezing it down in liquid nitrogen to embrittle it, and then crushing it to pyrophorically fine powder: DREADCO's 'Pyrubbidust' (Regd). This will be a splendid new energy resource. Kept under nitrogen, and floated through pipes on a stream of that gas, it will ignite as soon as mixed with air in a furnace. On a small scale, sachets of Pyrubbidust will make ideal matches and firelighters. Daedalus even has plans for a cigarette whose pyrophoric tip ignites when the air-excluding wrapping is torn off (his previous autocigarette, which you struck on the packet, never caught on). And self-heating cans of soup, campers' hot-water bottles, emergency flares, etc., can also be envisaged. Clearly a very simple Pyrubbidust internal combustion engine is possible, the powder being directly injected into the cylinders. No ignition-system would be needed, but there might be abrasion and disposal problems from the solid ash, which will have to be collected in a sort of vacuum-cleaner bag. Daedalus hopes that the ash will be even finer than the fuel, and like jewellers' rouge will merely smooth the inside of the engine to a high polish.

(*New Scientist*, 23 January 1975)

'A voidless solid board whose humble origins will be proclaimed by a fascinating surface grain . . .'

A knotty problem for Maxwell and Faraday

Faraday's 'lines of force' must be among the most satisfying of intellectual inventions. These elegant abstractions go from one pole of a magnet to another, cannot terminate or cross, and behave like elastic strings which mutually repel. They enable one to understand the whole qualitative behaviour of a magnetic field without mathematics, and suggest to Daedalus that fields of unusual subtlety may be possible. He proposes a novel experiment based on the old party trick of tying a knot in a piece of string without letting go of the ends (you grasp the ends with your arms folded, and then unfold your arms). His scheme is to wind a coil around a bar of flexible magnetic ferrite-rubber, and tie a knot in it. Energizing the coil with an electric current would give a knotted magnet. Undoing the knot would then, since lines of force cannot cross, transfer the knot to the magnetic field. (It may prove necessary to sheath the coil in superconducting metal to prevent the lines of force penetrating the material of the magnet; but this tedious cryogenic complication doesn't affect the principle of the thing.)

A knotted magnetic field is rather a subtle topological concept. Its main feature is that the lines of force, which tend to contract and pull the knot tight, are bunched together where they go through each other. The closer they pack the higher is the resultant field, so the net result is a region of convoluted but high magnetic field, surrounded by a region of lower and straightforward field. At last Earnshaw's theorem, which forbids a magnetic field to be stronger anywhere than it is at the surface of the magnet causing it, will be overcome! Truly stable magnetic suspensions will thus at last be possible. For a piece of iron dropped into a knotted magnetic field will not immediately fly to one of the poles. It will fly to the knot, the region of highest field, and there remain suspended. Delicate instruments, easy chairs, vehicle-springing and magnetic-suspension transport systems would all be revolutionized. Furthermore, the knots would interact in a curious fashion with current-carrying conductors, yielding electric motors whose moving parts would describe elaborate closed-cycle paths. Conversely, a plane conductor traversing a knotted magnetic field would have a knotted electric current induced in it, though even Daedalus has not thought of a use for this.

The interaction of two or more knotted fields will be even more subtle. Where the lines of force nearly close on each other, a magnetic knot should behave like a little segment of current-carrying conductor, i.e. an electric dipole. So knots should attract each other as dipoles do. This effect will be opposed by the repulsion of their parent lines of force, but may still permit their aggregation into rather a complex knitted magnetic field.

But Daedalus is most bemused by the fate of a knotted line of force when the current is switched off. Ordinary lines of force contract to a point and vanish, but he is not sure if this is feasible for a knot. It may shrink right down to the quantization-limit of the electromagnetic field, but be forced to persist as a tiny unit of electromagnetic energy. Daedalus may have invented the quark.

(*New Scientist,* 5 October 1967)

The DREADCO knotted magnetic suspension finds many uses in the home — keeping paper-clips handy, holding painted metal objects while they dry, etc. Here it provides a frictionless 3-D bearing for this full-movement pet-mouse exerciser

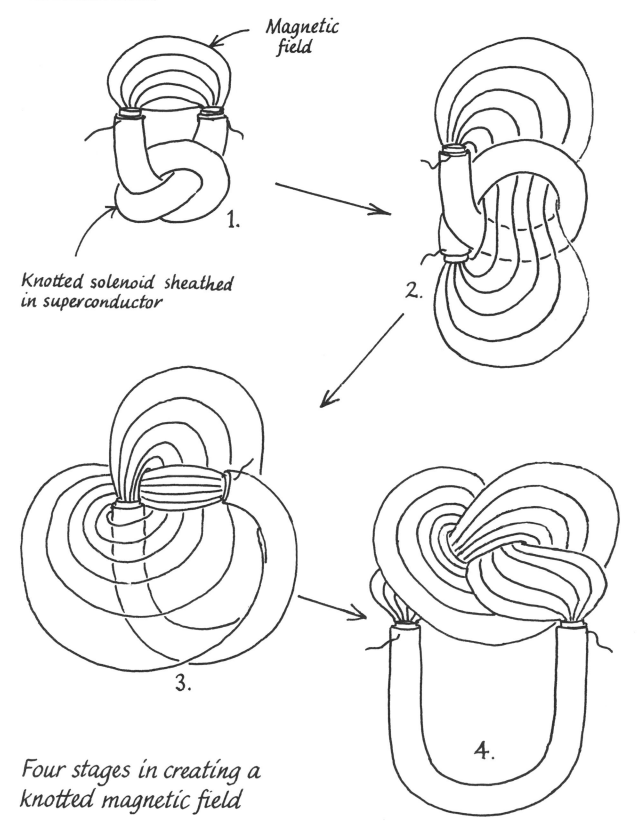

Magnetic field

Knotted solenoid sheathed in superconductor

1.

2.

3.

4.

Four stages in creating a knotted magnetic field

Fire in the belly, and where it gets you

Daedalus has been pondering the strange fact that cows, like many herbivores, evolve copious quantities of methane from their digestive tracts. He wonders how much methane those herbivorous dinosaurs must have produced — and is developing a theory that they met their end when the rising oxygen-content of the atmosphere created explosive-mixture conditions in their interiors. Unlike all other extinguished species, their world ended with a bang rather than a whimper. But if creative evolution is nudging the humble cow down the same track again, it must presumably have some worthwhile goal: so Daedalus is helping the process along. He is selecting the gassiest calves and feeding them on a high-cellulose (e.g. newspaper) diet to see what ultimately transpires. Seeking a natural source of ignition, he recalls the flickering 'Will-o'-the-wisp' phosphorescence in marshes. This is thought to be due to biodegradation of phosphorus-rich organic remains to spontaneously inflammable phosphine gas. So he is including plenty of phosphorus-rich fish-bones in his cattle-cake.

The first goal of this accelerated-evolution programme is a cow with an explosion-powered larynx, like those exploding acetylene-air bird-scarers. He reckons that a really deafening moo would enable the creatures to maintain herd contact over wide areas, and would also give them a novel defence against predators. If in addition their vocal blasts could propel chewed fodder at lethal velocities, they would become the first herbivores to carry defensive artillery. Such supercows would colonize the wildest jungle with confidence, opening vast new tracts to dairy farming. Trigger-happy rogue animals, however, might well provide the fastest-drawing cowboy with a novel challenge. Further evolutionary advances could yield the high-speed pulse-jet-propelled cow, though Daedalus cannot yet see how it would adapt to travelling backwards. And with really copious methane-generation, he suspects that a genuine, red, fire-breathing dragon will emerge — the ideal Welsh mascot!

(*New Scientist*, 13 May 1971)

Daedalus comments

The experience of fire-eaters who have backfired shows that a small-volume gas explosion in a wet-walled elastic enclosure (e.g. mouth or stomach) need not cause serious thermal or mechanical damage. But my remark about methane-propelled cows having to go backwards was needlessly restrictive. Methane is not only evolved from the creatures' rumen to be dis-

charged by belching. Stephen Pile, in his hilarious compilation *The Book of Heroic Failures* (Routledge & Kegan Paul, 1979) rescues from oblivion a news item of August 1977. It concerned a Dutch veterinary surgeon treating an ailing cow. 'To investigate its internal gases he inserted a tube into that end of the animal not capable of facial expression and struck a match. The jet of flame set fire first to some bales of hay and then to the whole farm, causing damage estimated at £45 000. . . . The cow escaped with shock.'

Of the million or so animal species known, over 250 000 are beetles. This heavy investment by Nature must betoken some ambitious evolutionary plans. So Daedalus has been studying the beetle tribe in search of clues, and is greatly intrigued by the striking defensive weaponry of the bombardier beetle. *Brachynus explodens* generates in its insides a strong solution of hydrogen peroxide, and in a separate compartment the enzyme catalase which catalyses its conversion to steam and oxygen. On being threatened, it mixes the two in a combustion chamber and, with a loud report, squirts the boiling mixture at its assailant. Though the resulting propulsive thrust is still quite small, Daedalus cannot avoid the conclusion that *Br. explodens* is well on the way to evolving the rocket-motor! The beetle's choice of hydrogen peroxide, the fuel of many man-made rockets, is significant. If it was merely evolving (say) a high-acceleration avoidance-manoeuvre to evade predators, an air-breathing combustion system would be more economical. Only for spaceflight is hydrogen peroxide, as a self-contained fuel, ideal. Many beetles can live for long periods at low pressures and without external oxygen; they can survive wide temperature variations, too. And a tiny, light beetle would find re-entry infinitely easier than does a 2-ton spacecraft. So Daedalus's natural, organic space-programme has begun a selective breeding of bombardier beetles to speed up Nature's long-term plan. Repeated selection of the fiercer squirters should soon produce specimens of considerable propulsive power. He is also encouraging them to become parasitic upon high-flying birds like godwits, as a convenient first stage. And for communication and telemetry, he is investigating the widely suspected phenomenon of telepathy in the amazingly well-organized communes of social insects like ants, bees, etc.

(*New Scientist*, 5 August 1971)

Daedalus comments

Professor J. F. Allen of the Physics Department of the University of St Andrews, wrote to *New Scientist* in protest against my space-beetle theory. He argued that even a single organic molecule must be decomposed by re-entry, so a beetle stands no chance at all. However, Professors Sir F. Hoyle and C. Wickramasinghe (*New Scientist*, 28 September 1978, p. 946) imply that virus particles at least can survive re-entry, as they reckon that influenza epidemics may start this way.

Indeed, they have even suggested that bacteria can re-enter safely, and that many interstellar dust-grains are really dormant bacteria drifting between planets. My own feeling is that creative evolution could easily cook up some safe mode of re-entry for aspiring beetles — repeated glancing re-entry interspersed with radiant cooling, use of the insect exoskeleton as a sacrificial heat-shield, burrowing into meteorites, etc. And if the beetle *were* the primary life-form, from whose early colonists all others have evolved, it would explain why there are so very many of them.

Inverting the managerial pyramid

The energy crisis has been taken to herald the beginnings of the transition to a steady-state, no-growth economy. Up to now, despite multiple technological revolutions, the consumption of energy has been rising at a steady 4.5% per annum for well over a century. This certainly compares remarkably closely with Professor Parkinson's figure of 5–6% per annum as the rate of bureaucratic expansion, independent of the work (if any) to be accomplished by the bureaucracy. Daedalus believes that it is Parkinson's Law which causes industrial expansion, by fostering the absurd career-prospects of the capitalist ethic. Thus if an industrial hierarchy has four underlings to each person on the next higher level, and everyone expects promotion every five years, then the firm must grow at a grotesque 32% per annum indefinitely. More realistically, assume a spread of talent. Just as 20% of the people drink 80% of the beer, so it is about true that 20% of the people do 80% of the work. Now people work their hardest on a challenge offering about half a chance of success: much higher or much lower chances sap their motivation. So if the ambitions of this active 20% drive the economy, then the growth-rate which maximizes their efforts should give each of them half a chance of promotion per 5-year period. This comes out at 5.4% per annum: gratifyingly close both to Parkinson's figure and to the growth in energy-consumption —which at 4.5% per annum is probably the best general indicator of economic growth. Notice that since the industrialized population has only been growing at about 1% per annum over the last century, some 3.5% of firms must go bust every year to provide new intake for the promotional population-growth of the rest. This looks about right too.

So, says Daedalus, the changeover to a steady-state economy can be brought about merely by changes in office-politics. If industrial concerns regrouped so as to have only two underlings per overling, the change in the hierarchical pyramid would by itself reduce the growth-rate to 2% per annum. And with a 1:1 ratio, giving each layer in the pyramid the same number of occupants as the next layer up, economic growth would cease. Each round of promotions would move everybody up one; the top layer would retire and a new bottom layer would be taken on. Everybody's ambitions would be satisfied with no need for the firm to expand at all; although (as Gilbert and Sullivan have pointed out) the satisfaction of being one of several hundred company chairmen is not as great as that of holding the same office unaccompanied. Even more extreme, modern industrial trends may even invert the pyramid. One worker at the fully automatic machine tool may need 4 managers to hammer out his computerized production schedule, and 16 executives to devise the deci-sion-matrix-based market strategy underlying it. In such a population-inversion, promotion will in fact cause company shrinkage! Daedalus concludes that the eco-nuts are exactly wrong. The way to reverse destructive economic expansion is not to revert to old-fashioned methods, but to push ahead with hypermodern automation as fast as possible. The one redeeming feature of pre-industrial society was its high death-rate. Enough people died off naturally to gratify the ambitions of the survivors, who could climb the steadily depleted pyramid without causing economic growth. So another answer to the ecological-industrial crisis may be to re-introduce duelling.

(*New Scientist*, 20 December 1973)

From Daedalus's notebook

Consider a hierarchy in which each level has t times as many members as the one above it; and consider level l with N members. The level above, $l + 1$, will have N/t members and the level below, $l - 1$, will have Nt members.

Now let a proportion p of all members be promoted. Level l is left with a residue $N(1 - p)$ of disappointed candidates, augmented by Ntp new arrivals from below. So its population is now $N[tp + (1 - p)]$: i.e. it has grown by a factor $F = tp + (1 - p) = p(t - 1) + 1$.

If promotions occur every 5 years, then the hierarchy will have an average annual growth-factor of $\sqrt[5]{F}$. In percentage terms, this is an annual increase of:

$$I = 100[\sqrt[5]{F} - 1]\%$$
$$= 100[\sqrt[5]{p(t - 1) + 1} - 1]\%$$

Examples: If $t = 4$ (each hierarch has four underlings) and everyone is unfailingly promoted every 5 years ($p = 1$), then $I = 32\%$ per annum. More realistically, if 20% of the workers are active and in line for promotion, and half of them get it at each 5-year interval, then $p = 0.1$ and $I = 5.4\%$ per annum, a growth-rate which fits the observed long-term trend quite well. If the pyramid tapers less sharply, e.g. $t = 2$, then $p = 0.1$ gives $I = 2\%$ per annum only. A pyramid with no taper at all, $t = 1$, locks I at 0%. And an *inverted* pyramid, with $t = 0.25$, gives $I = -1.5\%$, i.e. 1.5% annual shrinkage! I wonder where they all go? This is clearly the way forward.

Daedalus comments

How nostalgic these musings of a few years ago seem today! We've managed to achieve economic stagnation by the good old-fashioned route of recession. Still, maybe the calculation has a moral for the future, or the Japanese.

Plumbing the heights

Daedalus's recent invention of the 'thermal glidoon'*, an aircraft powered by the temperature-difference between the lower and the upper atmosphere, is clearly capable of generalization. A large fraction of the world's electrical power goes into air-conditioning, so direct methods of tapping the cold of high altitudes would be very valuable. A balloon shuttle-service, ascending with buoyant loads of ammonia gas and returning with dewar vessels of liquid ammonia at $-33\,°C$ could easily be set up. But a static and continuous system would perhaps be more practical. Since the temperature of the air drops by about 6.5 deg C per kilometre of altitude, a balloon 1–2 km up, carrying a heat exchanger and tethered by a pair of pipes carrying coolant, could well provide air-conditioning for a house or even an estate. Windy weather would remain a problem, however (extra guy-ropes would be needed for stability), and larger sizes would become increasingly cumbersome.

So for a really ambitious exploitation of atmospheric temperature-differences, Daedalus is working out ways of extracting energy from cold mountain-tops. One promising candidate is Mount Kenya, which rises to 17 000 feet from equatorial Africa, and has a temperature at its summit of about $-18\,°C$. It could easily support a large heat-exchanger at its summit, with pipes running up to it. A condensible vapour would be passed up to the heat-exchanger, where it would condense to liquid; this would run down the return pipe and boil at the bottom in a second heat-exchanger exposed to the heat of the low-altitude tropics. The station at the bottom would generate electricity from the boiling liquid in a conventional turbine installation, together with useful 'central cooling' from the latent heat thus absorbed. The exhaust vapour from the turbines would then travel up to the summit again. An entire tropical town could easily be both cooled and powered by such an installation; conversely, the summit heat-exchanger would provide welcome warmth for intrepid climbers.

The choice of working fluid for such systems is rather restricted. A low molecular weight is needed, or the weight of the tall upgoing column of vapour compresses the vapour at the bottom so much that its boiling-point is raised above local temperature, and it condenses. Ammonia scores here with its low molecular weight of 17; but it would need to be pressurized to 2.2 atmospheres to condense at the top of Mount Kenya at $-18\,°C$, which implies rather heavy-gauge pipework. Methylamine

*See p. 104.

seems about the best choice. On the summit of Mount Kenya it would condense under only 0.6 atm, very close to the atmospheric pressure at that height. Even better, its molecular weight of 31 is very close to the effective value for air of 29, so its change of pressure with height matches that of the atmosphere. Accordingly the internal pressure in a methylamine pipe would everywhere balance the external pressure, allowing very light-gauge pipe to be used. Even small-scale schemes, like one to power Fort William from the cold at the summit of Ben Nevis, might become economically attractive.

(*New Scientist*, 17 February 1972)

From Daedalus's notebook

For the scheme in the diagram to work, the vapour in the upcomer must condense under conditions p_h, T_h at the top, and boil under conditions p_0, T_0 at the bottom. T_h and T_0 are set by the atmosphere and we can't do anything about them. p_h and p_0 are set by working conditions and by the vapour. The heavier the vapour (the higher its molecular weight), the greater the compression of the vapour at the bottom, the higher the boiling-point at the bottom and the worse everything gets. In the limiting case, the liquid only *just* condenses at the top and only *just* boils at the bottom. This implies that anywhere in the column, the pressure of the vapour is just the saturated-vapour pressure that the liquid would exert at that temperature. What molecular weight of vapour brings this about?

In the ideal-gas approximation, a vapour has density $\rho = pm/RT$ where p is its pressure and m its molar mass. So consider a short section of pipe δh, filled with the vapour. If the pressure at the top of the section is p, the weight of the dense vapour will increase the pressure at the bottom to $p + \delta p$, where δp is given by the hydrostatic formula $\delta p = \rho g \delta h = (pm/RT)g\delta h$.

Now in the limiting case, the pressure exerted by the vapour at any height is equal to the saturated-vapour pressure (s.v.p.) that the liquid would exert at that height. So if at the top of the section the liquid has s.v.p. p and temperature T, then δh metres lower, where the temperature is $T + \delta T$, it must exert s.v.p. $p + \delta p$. The increase of s.v.p. of a liquid with temperature is quite well given by the Clausius–Clapeyron Equation: $\delta p = \lambda p \delta T/(RT^2)$, where λ is the latent heat of evaporation in J mol^{-1}. Equating our two expression for δp we get:

$$\lambda p \delta T/(RT^2) = (pm/RT)g\delta h$$

$$\text{whence } m = (\delta T/\delta h)\lambda/Tg \text{ kg mol}^{-1}$$

Now most liquids obey Troutons's approximate rule that $\lambda/T \simeq 92\,\text{J mol}^{-1}\text{K}^{-1}$, where T is the boiling-point (as it is everywhere in the equilibrated tube). So putting $(\delta T/\delta h)$, the fall of atmospheric temperature with height, equal to its standard value of $6.5 \times 10^{-3}\,\text{K m}^{-1}$, and $g = 9.81\,\text{m s}^{-2}$, we find:

$$m = 6.5 \times 10^{-3} \times 92/9.81$$

$$= 0.061\,\text{kg mol}^{-1};$$

$$\text{m.w.} = 61\,\text{g mol}^{-1}$$

So unless we are lucky with λ or the various non-idealities slurred over by this calculation, we are limited to working fluids of m.w. < 61.

Try it with various possible working fluids. Firstly, the scale-height of an ideal gas (the height for which the bottom pressure is increased by a factor e) is given in the uniform-temperature approximation by $H = RT/gm$, where m as before is the molar mass. So

(a) *Ammonia*, with m.w. = 17 has $m = 0.017$ and $H = 13\,600$ m. At the top of Mount Kenya at $-18\,°\text{C}$, ammonia would condense at 2.2 atm; so 5000 m lower its pressure would be $p = 2.2\,\exp(5000/13\,600) = 3.2$ atm. The boiling-point at this pressure is $-7\,°\text{C}$, so the descending liquid will happily boil into such a pressure at tropical ground temperatures. A pity all the pressures are so high, though.

(b) *Methylamine*, m.w. = 31, so $m = 0.031$ and $H = 7500$ m. At $-18\,°\text{C}$, it condenses at 0.6 atm, so 5000 m down it will be at $p = 0.6\,\exp(5000/7500) = 1.2$ atm at which its b.p. is $-5\,°\text{C}$. So it will boil happily at the bottom. Furthermore, since the atmospheric pressure at the top of Mount Kenya is 0.54 atm, while at the bottom it is 1 atm, the pressure inside the tube will be very close to the pressure outside throughout its whole length. The pipework will carry very little stress, and can be nice and light. This is the one to go for.

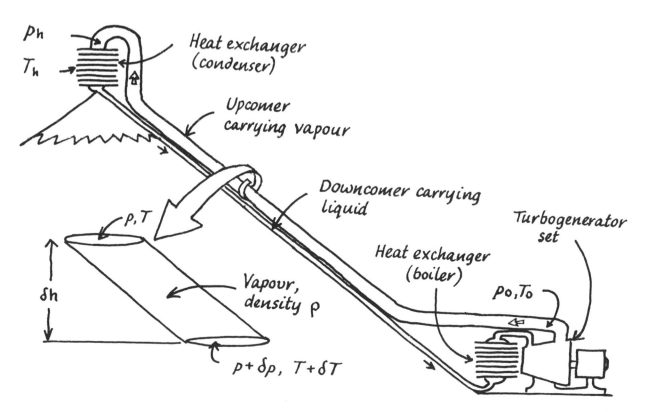

P_h
T_h

Heat exchanger (condenser)

Upcomer carrying vapour

Downcomer carrying liquid

p, T

δh

Vapour, density ρ

$p + \delta p, T + \delta T$

Heat exchanger (boiler)

Turbogenerator set

P_o, T_o

Ants and algorithms

The traditional engineering practice of estimating the maximum loading that a structure will have to bear, and then making it stronger than that by some 'safety factor', in effect pays a high price for ignorance. So Daedalus has been seeking some way of eliminating the guesswork, and the danger of unrecognized flaws in the structure, by measuring the actual stresses in the material and adjusting the structure accordingly. He recalls the intriguing habits of termites. These wood-eating creatures don't attack the surface of a wooden structure, but frequently riddle its insides with so many tunnels that the resulting hollow shell collapses at the slightest touch. From this observation Daedalus deduces that the canny creatures must sense the stresses in the wood as they chew it, and know when further inroads would lead to disastrous collapse. Now a termite can only digest wood by virtue of special microorganisms (*Trichonympha*) in its stomach. So Daedalus suggests replacing these by alternative species. In recent years, a whole new army of microbes has evolved to exploit the new materials provided by man's industry. There are microbes that chew up plastics, fungi that live on glass or attack aluminium, etc.; and these might well be persuaded to set up shop in the stomachs of termites. Airframes and other critical assemblies would then be stressed to the maximum required limits, and the termites would be released in them. All superfluous internal metal would be removed, leaving a 'skin-covered sponge' structure which would be the lightest possible framework for the job. Hidden weaknesses, etc., would be automatically allowed for by the stress-sensing of the insects. The only problem would be in preventing the creatures from escaping, for they could wreak havoc with much of the metal-work of civilization. Daedalus, however, recalls that when one warms termites up, their stomach-microbes die before they do, condemning them to indigestion. So he hopes to breed thermophobic strains of the metal-degrading microorganisms so that the termites could be annealed after use to render them harmless.

(*New Scientist*, 14 July 1966)

Daedalus is endlessly amazed by the subtle instincts of insects. The exact curve-tracing of the leaf-cutter beetle, the splendid webs of spiders, the elaborate architecture of termite nests, all show what can be done with blind instinct. This thought reminded Daedalus of the great imbalance in current human technology. Electronics can be incredibly miniaturized while mechanics is still very clumsy, and Daedalus hopes to recruit the mechanically gifted insect world to redress the balance. He recalls that irradiated spiders often spin very odd webs. The radiation must alter the software of their instinct, so he is irradiating the eggs of ants, spiders, wasps, etc., to look for mutated new instincts. Most of these will be bizarre and destructive, but the odd few lucky variants should be useful industrially. (Ultimately genetic engineering may be able to synthesize and implant the chemical software of new instincts far more elegantly.) The first goal will be a simple wire-connecting instinct enabling ants to make connections to integrated circuits — a task now laboriously conducted by human workers with microscopes and micromanipulators. More elaborate work could be carried out by teams of insects in collaboration, like wasps building a nest. But the constructional instincts of insects are basically very simple. They are algorithms (sets of rules) with only four or five separate instructions. The redesign of objects to be assembled or carved from the solid by appropriate ingenious instinctive rules will be fascinating.

Complex objects — a complete telephone, say — would be beyond any simple algorithm. Each subassembly or sub-subassembly would have to be made by its own insect team with its own specialized instincts; and the interconnection of these assemblies would need other teams still. By current design criteria, these biological products will be very strange. They will all be different, for a start; and there won't be a plane surface or a precise tolerance anywhere. They may be beautiful or grotesque or a bit of both; and yet the blind adaptive feedback of their construction will ensure that all their tiny delicate mechanisms unfailingly work. But if a fault does develop, they need only be replaced in the right area of the assembly line, and the industrious workers will put them right automatically!

(*New Scientist*, 26 February 1981)

Daedalus comments

Termites are notorious for weakening wooden structures from within so that they collapse unexpectedly under human loads. A splendid example is provided by the story of an Indian cricket match in which the stumps were unwisely left in the ground overnight. Soon after play was resumed the next day, a batsman was clean bowled — shattering the stumps to fine splinters! This ultimate achievement in demon bowling was the result of nocturnal sabotage by the local termites.

Philip Morrison gave a good example of a termite constructional algorithm in the 1979 Jacob Bronowski Memorial Lecture given at MIT, broadcast by the BBC, and subsequently printed in *The Listener* (23 August 1979, p. 234). According to Professor Morrison, termites elaborate a tacky mixture of wood-fibre and saliva, a sort of papier-mâché called 'carton'. They form it into pellets which mutually adhere and set solid. To build a termite-nest — an incredibly complicated structure which may be up to 20 feet high — all the insects follow this algorithm blindly:

1. Make a pile of pellets out of carton.
2. When the pile has reached a specified size, explore your vicinity to discover if it contains a bigger pile. If so, abandon yours and work instead on the bigger one.
3. When your pile has reached another, bigger, specified size, explore your vicinity to discover if it contains a neighbouring pile close enough to be bridged to your pile. If not, abandon your pile and search for another which does have such a neighbour, and work on that.
4. If your pile does have a neighbour close enough to be bridged, expand the top of your pile to join the two in an arch. Then carry on as before.

The result of thousands of unorganized insects all blindly following this algorithm is the elaborately tiered and arched structure which is a termite nest. Nowhere is there any plan or blueprint, and no two nests are alike, but the structures always work as nests. There must be subsidiary rules in the algorithm to cope with the outer roofing-over of the structure, and with its magnetic orientation (for termite nests often show an intriguing north–south directionality) but even so the algorithm is orders of magnitude simpler than the sketchiest of blueprints. The design of objects to be constructed by such algorithms, and the algorithms to do it, should be a compelling intellectual discipline.

(Courtesy of *New Scientist*)

Aquahovercraft

The long history of misdesign and aesthetic disaster in modern architectural and town-planning practice suggests to Daedalus that buildings should be made mobile, so that repeated experimental redevelopments could take place without vast efforts of demolition and reconstruction. The hovercraft principle seems ideal for this purpose, but since buildings typically exert from 0.02 to 2 atmospheres pressure on the ground beneath them, it could be hard to generate the required lift. And the noise and draught would be fearful. Daedalus therefore intends to use water, a thousand times denser than air, as the lifting medium. An 'aquahovercraft' pumping water out beneath itself could provide really massive lift at quite modest flow-rates. Unfortunately it would also flood the neighbourhood unless the emitted water were contained somehow. Daedalus's aquahovercraft system incorporates a suction pontoon (a scaled-up version of that little gadget of the dentist's) surrounding the craft to capture and recycle the emitted water. He is designing buildings incorporating all the water-tanks, pumps, retractable skirt mechanisms and suction-pontoons needed to make them aqua-mobile on demand. Many existing buildings, especially those built on shallow foundations and on the concrete-raft principle rather than on deep piles, could also be 'retro-fitted' with the system.

Gushing ponderously, these urban juggernauts will steam to and fro at the whim of architectural fashion: high-rise flats may move out while sections of elevated motorway move in, and little shops and semis scurry around their feet. Factories will tour the country in search of skilled labour or development grants; empty office-blocks will creep towards central London to increase their value while full ones, urged by the relocation bureau, edge out. Black ghetto housing and while suburban property could be juxtaposed or segregated according to current policy, or left to reach their natural equilibrium: in all sociological fields the planners could at last learn from and even rectify their mistakes. Only the map-makers and piped-service engineers would bemoan the new Utopia.

(*New Scientist*, 3 February 1972)

From Daedalus's notebook

Consider a hovercraft of radius r, ejecting fluid radially outwards between the ground and its skirt through a gap of width x. The total area of the gap is $A = 2\pi rx$, so if the fluid has velocity v and density ρ, the total mass-flow per second is $\dot{m} = Av\rho = 2\pi rxv\rho$.

The pressure p induced upstream of this outflow must equal the momentum-flux per unit area of aperture, i.e. mass-flow × velocity/area:

$$p = 2\pi rxv^2\rho/2\pi rx$$
$$= v^2\rho$$

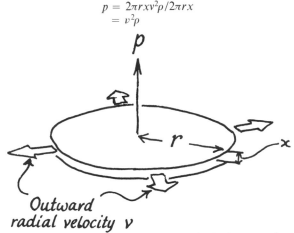

Outward radial velocity v

This pressure exists everywhere inside the hovercraft, and acts upwards on the cover plate radius r. So F, the total lift exerted, is the pressure times that area:

$$F = \pi r^2 v^2 \rho$$

Clearly water, with ρ a thousandfold greater than that of air, increases the lift proportionately. With reasonable values for a building-lifting hovercraft: $r = 10\,\text{m}$, $v = 10\,\text{m s}^{-1}$, ρ for water $= 1000\,\text{kg m}^{-3}$, we find:

$$p = 10^2 \times 1000\,\text{N m}^{-2} = 1\text{ atm};$$
$$F = \pi \times 10^2 \times 10^2 \times 1000$$
$$= 3.1 \times 10^7\,\text{N}$$
$$= 3100\text{ tons-force}$$

Clearly OK.

What power will be needed to maintain the flow? With a tough elastomeric skirt hugging the ground contour and limiting the gap x to (say) 1 mm, the mass-flow will be $\dot{m} = 2\pi r x v \rho = 2\pi \times 10 \times 0.001 \times 10 \times 1000 = 630 \, \mathrm{kg \, s^{-1}}$. So the power in the water-flow $P = \frac{1}{2}\dot{m}v^2 = \frac{1}{2} \times 630 \times 10^2 = 31 \, \mathrm{kW}$; about 40 horsepower. Not at all excessive. And everything gets better as the aquahovercraft gets bigger, for lift rises with r^2 while mass-flow and power rise only with r.

Daedalus comments

This calculation was so straightforward as to be positively worrying. So I was not very surprised to dis-cover subsequently that other organizations were hard on the heels of DREADCO. Six months later (*New Scientist,* 17 August 1972, p. 340) came a report that the National Engineering Laboratory in East Kilbride was using 'hoverwater' pads to move heavy loads around in shipyards. The patents were being handled by the National Research and Development Corporation and I wonder if any aspect of them could be invalidated by prior publication? The main application envisaged was the accurate positioning of heavy castings for assembly. But so far, nobody else seems to have suggested using the system for manouevring whole buildings.

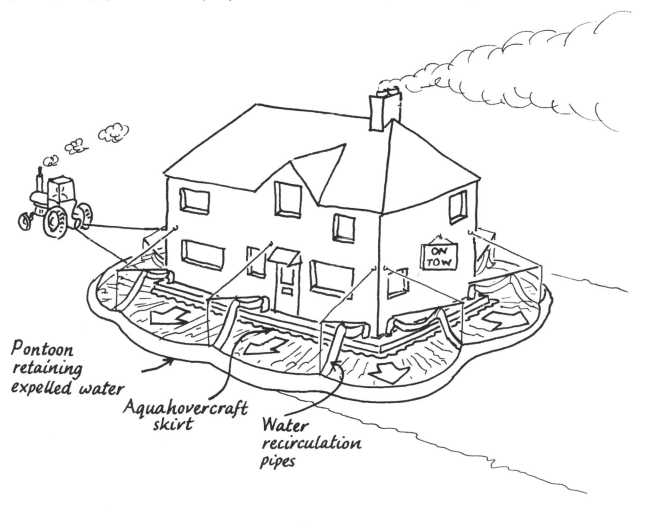

Pontoon retaining expelled water

Aquahovercraft skirt

Water recirculation pipes

Nastyglass

Apart from pain, it is not obvious that there are any inherently unpleasant sensations. Is the smell of hydrogen sulphide or the sight of a spider inherently offensive, or do we learn to dislike these things? Recently, however, the audio industry seems to have come across an inherently unpleasant sound: the 'crossover distortion' of certain transistor amplifiers in which a sine-wave signal falters irregularly as it crosses the axis between positive and negative peaks. Such a distorted signal has an annoyance-value out of all proportion to its content of added harmonics; the ear, long acclimatized to natural, near-sinusoidal sound, seems severely pained by this alien waveform. Daedalus advocates adding crossover distortion to the voices of villains in radio or TV drama for extra repulsiveness, and crossover-sabotaging the oilier ads and party political broadcasts. But more intriguingly, he is now extending the principle to light, another naturally sinusoidal waveform. The new science of non-linear optics is revealing many materials whose optical properties vary with the instantaneous strength of the a.c. field which is light. So DREADCO physicists are seeking an Ovshinsky-type glass which is conducting (and hence opaque) at low light-fields but becomes insulating (and therefore transparent) at high ones. Such a glass would neatly clip out the low-intensity regions where the light waveform was crossing its axis, while passing the peaks with little attenuation. In other words it would impose on the light the most severe crossover distortion.

Daedalus cannot predict how things will look seen through DREADCO's 'Nastyglass' (Regd) but he imagines they will have the leering, sickly quality of severe 'morning-after' light. Colours will only be slightly distorted by the added harmonics but in a clashing, fatiguing, headachey manner. Daedalus envisages many applications in traffic lights, warning signs, and the more way-out of picture galleries. But the main outlet will surely be as a tapering-off aid to the TV addict. A Nastyglass screen-cover could nudge many a hopelessly-hooked square-eyed junkie onto the long road back to sanity and mental independence.

(*New Scientist*, 22 July 1971)

From Daedalus's notebook

With the old valve amplifiers, fidelity was well measured by 'total harmonic distortion', which was imperceptible if below about 0.2%. But many transistor amplifiers manage to sound quite horrid despite an equally low t.h.d. A prominent school of the audio fraternity (e.g. R. Williamson, *Wireless World*, Vol. 75, June 1969, Amplifier Supplement) blames crossover distortion. This is a vice of class-B amplifiers in which different transistors handle the positive-going and the negative-going peaks. If the transfer characteristic between them is not exactly smooth, a sine-wave input gets 'crossover-distorted'. The effect is painful out of all proportion to the introduced harmonics (Diag. 1).

Can we work the same trick with light? The obvious starting-materials here are the Ovshinsky semiconductor glasses. They can have various desperately non-linear characteristics, and the one we want is the current-limited form (Diag. 2). Consider what a pane of this glass would do to the electric vector of light. At low fields (between points A and B) each increase of electric field increases the current drawn by the glass. It acts as a resistive conductor, dissipating or reflecting the incident energy just as conducting metals do: so it is opaque. But at fields higher than B, no further current is drawn no matter how high the field goes. The glass is practically an insulator and should transmit the incident energy almost without loss. The net result should be that the low-field sections of the original sine-wave will be clipped right out: crossover distortion! Formally, this is equivalent to adding various higher harmonics to the light, but with any luck the effect on the eye — which is by no means a spectrometer — will be far worse.

Uses for nastyglass. Various psychological conditioning and masochistic art-form applications come immediately to mind. But there may be simpler ones. A crossover-distorting audio amplifier often, paradoxically, sounds far worse at low volumes than at high ones (the fixed 'glitch' on the signal as it crosses the axis is a much smaller proportion of the total output at high power). Analogously, Nastyglass should be far worse in dim artificial light than in bright daylight. Indeed, at sufficiently low light levels the electric vector will never get outside A–B and the glass will remain opaque throughout the whole cycle. So Nastyglass windows could replace curtains. They'd let the bright daylight in with little degradation; but at night they'd completely retain the dimmer artificial light within the room. And any that did get out would be so horrid that no Peeping Tom could bear to look at it.

1. Crossover distortion in an audio amplifier

Input signal   Output signal

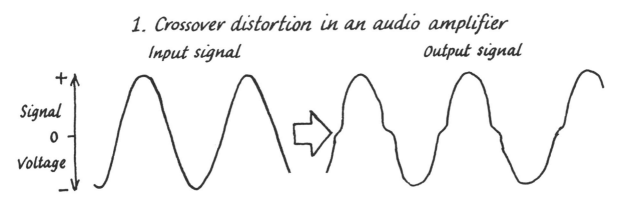

2. Crossover distortion in an Ovshinsky glass

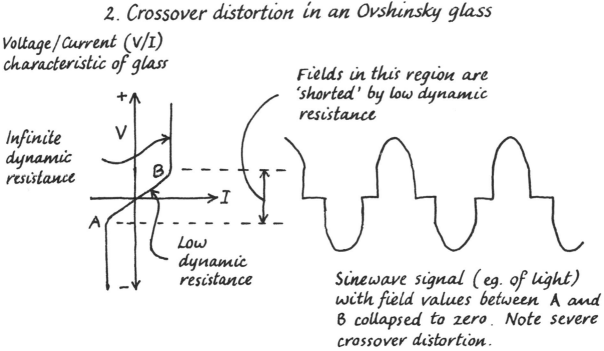

Voltage/Current (V/I) characteristic of glass

Infinite dynamic resistance

Low dynamic resistance

Fields in this region are 'shorted' by low dynamic resistance

Sinewave signal (eg. of light) with field values between A and B collapsed to zero. Note severe crossover distortion.

Boredom in the blood

Current researches on biological clocks in animals and man include some novel investigations by Daedalus. He points out that, since time races past while we are engrossed, but drags along when spent boringly, a disproportionate amount of our subjective life is devoted to tedium. Like so many other mental effects, this distressing phenomenon is probably governed by some substance secreted into the blood, in this case a 'time-dilator'. So Daedalus aims to isolate it, and use it to liberate humanity from the shackles of subjective time.

His first idea was to design a 'vampire collection-plate' which, when passed among the congregation after an interminably boring sermon, accumulated blood-samples from the weary fingers via an anaesthetizing needle. But most mental agents work in such tiny doses that a more intensive approach is indicated. So Daedalus is now collaborating with avant-garde artists in staging the ultimate marathon bore-in. The event will draw heavily on his own experiences at scientific conferences. A series of seemingly endless technical speeches will be given by 'foreign dignitaries' whose eminence and earnestness will inhibit the audience from guilty defection or open contempt. About 30% of their thickly accented words will be incomprehensible, and their attempts to maintain human interest via atrociously unfunny and long-winded personal anecdotes will introduce a valuable additional element of embarrassment. The talks will be illustrated by irrelevant and tantalizingly out-of-focus slides interspersed with subliminal flashes suggesting to the viewers that they have left the kettle on or the bath-water running. Sleep will be inhibited by subtly uncomfortable chairs, very high humidity and the absence of toilet facilities. These tactics should produce such excruciating time-dilation in the victims that the compound responsible should be easily discernible in their blood. Once identified it could be synthesized and produced as tablets to spin out the activities of real worth: to extend the delights of love and sociability and push back all the deadlines that threaten our hard-pressed race. The suffering inflicted on the experimental subjects would be well worth while.

(*New Scientist*, 18 July 1968)

Daedalus comments

> When as a child I laughed and wept, Time crept.
> When as a youth I waxed more bold, Time *strolled*.
> When I became a full-grown man, Time RAN.
> When older still I daily grew, Time *FLEW*.
> Soon I shall find, in passing on, Time *gone*.
> Oh Christ! wilt Thou have saved me then? Amen.

This poem by Henry Twells, which adorns the front of a clock-case in the north transept of Chester Cathedral, suggests that the time-dilating metabolite is present at high concentrations in the blood of children, but gradually diminishes in concentration as life proceeds. This may prove a useful clue to its identification. Furthermore, the educational establishment has developed to a very high level the various techniques for keeping children bored, so a programme of blood-sampling from the more formal of pedagogic institutions might also be rewarding.

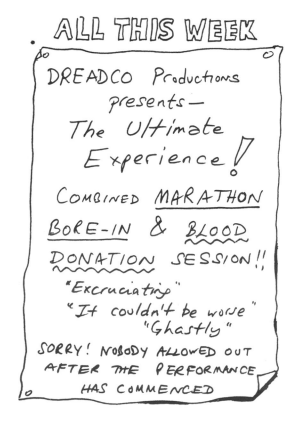

ALL THIS WEEK

DREADCO Productions presents —
The Ultimate Experience!

COMBINED MARATHON BORE-IN & BLOOD DONATION SESSION!!

"Excruciating"
"It couldn't be worse"
"Ghastly"

SORRY! NOBODY ALLOWED OUT AFTER THE PERFORMANCE HAS COMMENCED

Crackers

Conventional methods of cutting, sawing, and shaping things are incredibly wasteful of energy. A lathe, for example, is much less than 0.1% efficient — most of its power appears uselessly as heat in the metal-cuttings. The most efficient cutting process that Daedalus can think of is glass-cutting. A glazier can draw a line on the glass and snap it to crack cleanly along that line, with very little expenditure of energy. The process is very flexible, too: in skilled hands it can make wavy shapes and even cut holes. Once initiated, cracks can also spread in metals (ships sometimes just snap in two on the high seas) and the process is greatly aided by low temperatures. So DREADCO technologists are now developing 'crack technology' to replace clumsy sawing and machining. Once initiated, a crack runs at several kilometres a second. To control it to less than a millimetre, the cracking system must react in under a microsecond: quite leisurely by electronic standards. The prototype 'Dreadcracker' pre-stresses the workpiece by fast-acting piezoelectric transducers, initiates a crack from a filed notch at the edge, and monitors its progress by photo-electric and strain-gauge signals. Just as one can steer a tear through a newspaper by careful differential pulling on the sides of the tear, so the Dreadcracker registers the progress of the racing crack and keeps it on program by signals fed back to the stressing transducers. It all happens in milliseconds: the machine operator just sees the workpiece magically crack into the chosen pattern. Any shape can be pre-programmed for the cracker. It would be tricky, but possible in principle, to take a large painted ceramic tile and shatter it into a precisely-designed jig-saw puzzle of interlocking parts!

For three-dimensional shaping a sheet-crack must be propagated, but the principle is the same. The DREADCO workers are developing the process using materials conveniently brittle at room temperature: thus they are seeking to crack instant lenses out of a glass block to sidestep all that grinding and polishing. When the process is well worked out, it will be used on metals embrittled by low temperatures. Thus Daedalus hopes ultimately to form a complete cylinder block in one hectic cracking of a chilled iron casting. The great attraction of the process, besides speed and energy-economy, is the absence of waste. In this case the cylinder centres and hole-cores, contracted by the release of stress, will just slide out of the crack-formed holes, and will be available for economic use in making bolts, pistons, etc. for the same engine.

(*New Scientist*, 22 April 1976)

The thing engineers have nightmares about is brittle fracture. A small crack in a highly stressed region of a component can, once started, run right through the whole thing with disastrous consequences. Daedalus has been musing that our neolithic ancestors were the real masters of brittle-fracture technology. They developed flint-chipping to a fine art, unknowingly exploiting the subtle fact that most flints have a surface layer in which the speed of sound is less than that of bulk flint. Now the speed of a crack is governed by the speed of sound in the solid. So a crack started obliquely from the surface of a flint is 'refracted' by this surface layer and just comes out again — detaching a neat flake rather than shattering the flint. So, says Daedalus, let us revive Flintstones technology! Many surface treatments for metals (e.g. nitriding and carburizing) diffuse alien atoms into the surface to change its properties. Daedalus is seeking a treatment to give very low surface speed-of-sound, so that any crack will travel a sharply curved path. The resulting crack-proof metal will revolutionize engineering. Imagine a component being loaded in a test-rig (or even on the job). When any point reaches a dangerous stress-level, the resulting crack will not spread through the material. It will be refracted back to the surface, neatly removing the flaw or badly-designed stress-concentrating feature as a detached chip, and leaving a smooth curved surface beneath which the load will be distributed more evenly. Being unable to fail by brittle fracture, such components could be loaded up to the point where they fail by plastic flow. Now this often works *for* the engineer, enabling a badly designed part to yield slightly, thickening and work-hardening as it does so, and sharing its unfair burden with less loaded members nearby. So Daedalus's new metal will be practically self-designing! Daedalus envisages a new era of engineering in which designers are out of a job, things can be roughly slung together to optimize in service, and motorists view a trail of detached fragments behind their car with quiet satisfaction.

(*New Scientist*, 23 February 1978)

Right. An example of Daedalus's crack-deflecting mechanism in action. This lintel has begun to crack at the bottom under the weight of the masonry above it. Some chance in-homogeneity in the stone has deflected the crack through 90°; as a result it has failed to propagate any further, and the lintel has not split in two

One of the most remarkable examples of brittle fracture spreading from an initiating sharp notch occurred on the high seas between the Wars. In the fierce competition for transatlantic passenger traffic prior to World War I, the German shipping line Hamburg-America built the largest liner in the world, the *Bismarck* of nearly 57 000 tons. After World War I, as part of Germany's war reparations, she was handed over to the British White Star Line, and re-named the *Majestic*. The new owners made various improvements to the vessel, in keeping with their ideas of how a splendid and luxurious liner should be appointed. In 1928 they installed a new passenger lift. A succession of rectangular holes was cut through the various decks to receive the vertical shaft. During a subsequent transatlantic voyage, a crack started from the sharp corner of one of these holes, ran to the side of the ship, and proceeded down the ship's side until, by amazing good fortune, it ran into a porthole and stopped. In the highest traditions of the service, the matter was hushed up. Neither the press nor the 3000 passengers were told how close the biggest ship in the world had come to breaking in two.

The collective-responsibility vehicle

The economies of scale gained by large buses are counterbalanced by the inconvenience of waiting for them. Smaller but more frequent buses would be preferable, but for the expense of manning them. Daedalus recalls the supermarket principle — let the customer do for himself what a worker once did for him — and suggests that the passengers drive the bus themselves. His 'Collective-Responsibility Vehicle' has a steering wheel, controls, and TV view of the road ahead for every seat; and each passenger is invited to help drive if he can. A central mini-computer scans the signals from each steering wheel, accelerator, etc. It discards the most extreme values, on the grounds that they must come from the most inexpert or eccentric drivers, and averages the rest for transmission to the traction unit. Thus individual aberrations (the road-hog, or the man who wants to haul the bus off-route to his own doorstep) have no effect, but the mass knowledge of driving and of the bus-route are pooled. Of course, if everybody wants the bus to take an unofficial route, democracy wins, as it should.

Daedalus expects a bus driven in this way to have a solidity of purpose, a lack of individual style, which would make it very safe. Even a crew of drunks might have collective sense equal to one sober citizen. Problems might arise when one half of the passengers wanted to go right and the other half wanted to go left; but even if one faction did not give way as crisis approached, the computer could easily be programmed to recognize such dangerous bimodal distributions, and make a casting vote. The popularity of fruit-machines and fairground racing-car simulators suggests that the passengers might even pay for the privilege of driving, thus eliminating the conductor too. One promising approach would be to give each driving-seat a coin-box, and weight each passenger's driving signals by the money he inserted. Those who contributed most would dominate the vehicle's speed and style and route, a perfect expression of the sound capitalist principle whereby what happens to everybody depends on those who back their preferences with the hardest cash. The ensuing 'auction', as rival factions rammed money into the slots to take the bus their way, might be immensely profitable. Lord Rothschild, who advocates exactly this principle to govern the direction of scientific research, should be delighted.

(*New Scientist*, 29 November 1973)

From Daedalus's notebook

(a) *What driving characteristics will the Collective-Responsibility Vehicle (CRV) have?* (1) Enhanced accuracy. The signal/noise ratio of N summed observations improves by $\sqrt{N}$, so a bus with 64 passengers all driving will be controlled about 8 times as accurately. Even if they're all drunk, results may still be very sober. (2) Resolution of distributed values. A typical pattern of signals from the controls might be:

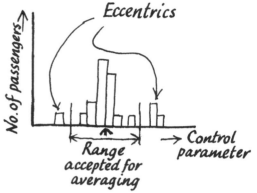

The computer 'truncates' eccentric values and averages the remainder. The outcome will be even safer because all the passengers are made aware of their collective choice by what the bus actually does. It's exactly like the 'Delphi' technique of prediction, where you ask a lot of experts each to make a private guess, then show them all the distribution of the guesses and ask each of them to make a second guess in the light of the first round. The process converges very rapidly to a consensus. Similarly the CRV should be very calm and certain in its driving style. (3) Multimodal distributions. Suppose half the passengers want to overtake and half want to stay in lane. The computer will be faced with:

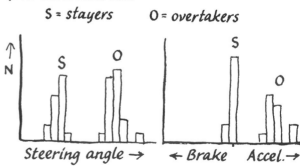

With such distributions, the program must ignore or truncate one entire block, i.e. make a casting vote. This must be done on the passenger rather than the signal level: i.e. one group of passengers must have all their signals ignored. Truncation purely on the signal level might result in the computer selecting the accelerator values of the 'overtakers' and the steering of the 'stayers', with disastrous results. A passenger who sees himself thus outvoted will first swing the wheel even more furiously — with no result as he is already truncated — and will then give up and return to optimizing what the CRV is actually doing, when his signals will be readmitted to the consensus. Again, the outcome should be very sane and stable.

(b) *How to organize the logistics of a CRV bus network?* The simplest method is to have no bus routes at all and let all the buses go wherever the passengers take them. To prevent permanent hijack by individuals we need (1) many bus-stops dotted around, each with some radiated signal that will always stop a passing bus and let new people on, and (2) an outer boundary, again electronically delimited, beyond which a bus can't be driven. It will stop there until new passengers come along to take it elsewhere within the area of service. With this system, the buses

would follow consensus-routes during rush-hours, when most passengers are going the same way; at other times they'd be a sort of flexible collective taxi service.

(c) *Payment for access to control.* This opens up some fascinating game-theory. In the rush-hour when everyone wants to go the same way, one paying 'delegate' could in theory drive for everybody. But since he could be outvoted by a single dissenter, who if unchecked could take the whole bus-load miles out of their way, all passengers have some incentive to chip in a little 'insurance money'. In less ordered circumstances it would probably pay to travel free and let the others drive as long as the bus was going roughly your way. But as soon as it took a seriously wrong turning you would either (1) sink all your money at once to haul it back or (2) get off at the next stop and wait for another bus. Folk-routes hallowed by tradition would probably arise to temper with predictability the opportunistic randomness of the service. But payment-for-control would probably appeal best in excitable Latin regions (Rome, Rio de Janeiro). British phlegm would prefer fixed routes and a fixed entrance fee to work an on-board turnstile giving admittance to the bus.

Arachnautics

Young spiders can migrate quite long distances by spinning out a thread of gossamer which catches in the wind and carries them away. A long fine fibre of this sort makes a splendid spider parachute —an ingenious use of the principle that the viscous drag of a fibre in a gas stream increases with its length, but doesn't depend very much on its diameter. Daedalus calculates that a similar parachute for a man would need about 10 000 km of fibre. A single length wouldn't do — for one thing it would snap and for another the atmosphere isn't high enough. But 10 000 or so each 1 km long should work nicely, and if made of the finest possible glass or carbon fibre need only be 0.01 mm in diameter, giving a total weight of some 2 kg. Daedalus did wonder if a dedicated hippie could grow enough hair on his head to enable him to leap safely from buildings or aircraft in flight; but even a thick 200 000-hair thatch with every hair a metre long is hardly adequate. Besides, the hair is packed too densely for each fibre to be fully exposed to the airstream. Daedalus's fibre-chute will deploy an extending framework from which the filaments can spread out freely as they unwind.

This novel parachute will have many advantages. Not only will it slow its user gently as its fibres extend, instead of jerking brutally open, but it will be entirely invisible from the ground. The military would be able to drop parachutists relatively inconspicuously, and in any case the sight of seemingly unsupported personnel drifting down from the sky might demoralize the opposition. Daedalus also sees uses in aircraft-safety. A big parachute with its disastrous opening jerk cannot be used to save a stricken plane: but a mere ton or so of fibres payed out from a 70-ton craft could gently slow it to a few metres per second falling speed. And on the personal scale, Daedalus is devising a sporting fibre-pack which releases fibres or even melt-extrudes them spider-fashion. If released into the base of a thermal, the filaments would climb in the rising air and lift the wearer joyfully into the sky, later depositing him safely miles away. This new sport, combining the delights of gliding and ballooning, should perhaps be called arachnautics.

(*New Scientist*, 17 July 1975)

From Daedalus's notebook

The fibre as a parachute. What force is needed to haul a fibre through a viscous fluid? A good place to start is the standard formula for the shearing of a fluid between two long concentric cylinders, the inner one being hauled axially along at velocity v_0. The fluid velocity v_r at distance r from the centre is:

$$v_r = (v_0/\ln \gamma)(\ln r - \ln R)$$

where $\gamma = r_0/R$ is the ratio of the radii of the inner and outer cylinders. Differentiating with respect to r we get:

$$dv_r/dr = v_0/(r \ln \gamma)$$

This is the velocity-gradient in the sleeve-like cylindrical fluid layer between radii r and $r + dr$. Newton's law of viscosity then gives the shear-stress N resulting from this velocity-gradient:

$$N = -\eta \, dv_r/dr = -\eta v_0/(r \ln \gamma)$$

where η is the viscosity of the fluid.

This stress exists over the whole area $A = 2\pi rl$ of the cylindrical layer, radius r and length l, being sheared. So the total force transferred across the layer by the shear is:

$$\begin{aligned} F &= N/A \\ &= -2\pi rl\eta v_0/(r \ln \gamma) \\ &= -2\pi l\eta v_0/\ln \gamma \end{aligned}$$

r has vanished, as it should. The same haulage force F is handed from the inner moving cylinder on through all the layers of fluid to the outer fixed cylinder.

We now consider that this force results from the pull of gravity on a mass m attached to the inner cylinder — a spider, say. Putting $F = mg$ and rearranging, we find the velocity of fall to be:

$$v_0 = -mg \ln \gamma/(2\pi l\eta)$$

Now we imagine the outer cylinder expanding to infinity, leaving the inner one as a fibre of radius r_0 in free axial fall. But as this happens, the ratio $\gamma = r_0/R$ goes to zero, $\ln \gamma$ goes to $-\infty$, and v_0, the velocity of fall, becomes infinite! Something's gone wrong somewhere. Since in practice gossamer-suspended spiders don't fall and break their little necks, this tragic influence of the remote atmosphere can't be taken too seriously. The sensible way out is to assume some characteristic distance beyond which the air is not affected by the fibre, and call it R. The geometric mean of l and r_0, $\sqrt{lr_0}$, seems as good a choice as any. It only enters as its logarithm, so even if it's madly wrong it won't cause much error:

$$v_0 = -mg\ln(r_0/\sqrt{lr_0})/(2\pi l\eta)$$
$$= mg\ln(l/r_0)/(4\pi l\eta) \qquad (1)$$

Man-carrying fibre bundles. Suppose we have a 70-kg man, who can carry a parachute containing at most (say) $m = 2$ kg of fibres. If these are glass fibres of density $\rho = 2700$ kg m$_{-3}$, and spun as fine as possible, say radius $r = 0.005$ mm, their total length would be:

$$L_{tot} = m/\rho\pi r^2$$
$$L_{tot} = 2/[2700 \times \pi \times (5 \times 10^{-6})^2]$$
$$= 10^7 \text{m}$$

Let's divide this length into 10 000 fibres each 1 km long, and use the bundle as a parachute. The viscosity of air is 1.8×10^{-5} N s m^{-2} at 20 °C, each fibre is supporting 70/10 000 kg, so from (1) the parachutist's terminal velocity will be:

$$v_0 = \frac{(70/10\,000)\times 10 \times \ln[1000/(5 \times 10^{-6})]}{4 \times \pi \times 1000 \times 1.8 \times 10^{-5}}$$
$$= 5.9 \text{ m s}^{-1}$$

This is equivalent to jumping from a height of about 2 m, and is well within standard parachute practice. Very nice. Similarly, scaling everything up a thousand-fold, a 70-ton plane with 2 tons of fibres could be gently slowed to the same velocity. It's an attractive notion to imagine melt-spinning the fibres directly into the slipstream from spinnerets: the aerodynamic drag would probably draw them far finer than the best conventional fibre technology!

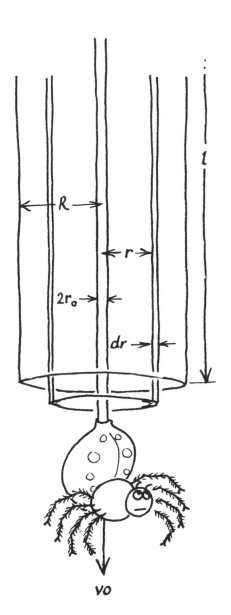

Per funicula ad astra

Daedalus, who started the whole aerospace business, has not lost interest in the field. He has been considering alternatives to launching satellites by rocket, which he thinks inelegant and wasteful. He currently has a scheme to erect on the Equator a tower 22 300 miles high. Such a tower, rotating with the Earth, would move with orbital velocity at its top; so you merely carry your satellite up and push it off. In case of opposition he has a cheaper plan which uses a single rocket to launch a satellite into a slightly higher orbit than this, while paying out say 24 000 miles of cable. The cable would then be anchored on the Equator and the satellite would hold it taut. Further small satellites could then be hoisted up the cable. Regrettably, Coriolis forces would tend to make the whole thing lag behind the Earth's rotation during this operation, but Daedalus reckons that ultimately the cable tension would bring it back ready for the next launch.

(*New Scientist*, 24 December 1964)

Encouraged by the lack of opposition to an earlier scheme of his to anchor a synchronous satellite with 24 000 miles of cable, Daedalus has come up with an even more ambitious project. He aims to establish a lift to the Moon. This would only require the launching of a rather bigger rocket carrying 10 times the previous length of cable and fitted with a harpoon head. The impact of such a rocket should anchor the cable firmly to the lunar surface and, since the Moon does not rotate seen from the Earth, no difficulty could arise at that end. Regrettably there are snags at the earthward end. The cable might be attached to a tower at the South Pole. Were it anchored to a point on the Equator, the Earth's rotation would wind the Moon in. Daedalus thinks that this result, while attractive as a geophysical experiment and short-cut to lunar exploration, might bring opposition from the opponents of Big Science. If, however, it is agreed upon, he suggests dropping the Moon in the Pacific, and from its goodness of fit to test the theory that this ocean is the crater left when the Moon was originally torn from the Earth.

(*New Scientist*, 16 September 1965)

Daedalus comments

These two jerky little proposals were among the very earliest of submitted 'Daedaluses'. Their rather apologetic tone, the expressions of regret and fears of opposition, arose because *New Scientist* at the time was still very worried about letting such bold suggestions loose in its respectable columns. So I was highly delighted when the key ideas of both these items were later proposed by J. D. Isaacs, A. C. Vine, H. Bradner and G. E. Backus in *Science* (Vol. 151, 11 February 1966, p. 682). These authors explained:

In addition to its self-support, such a cable installed near the equator of a rotating planet or natural satellite (or, in some cases, at the pole of a rapidly revolving body) and extending sufficiently beyond the radius of orbit of a synchronous satellite would have some other interesting and useful properties.

Masses that were moved along the cable from the surface of the central body would be launched into space with a release of net energy derived from the kinetic energy of the central body.

They went on to analyse the strength and taper of the cable and their dependence on the material of its construction, and to consider what planets and satellites of the Solar System lend themselves most readily to such schemes.

A subsequent correspondence I had with Arthur C. Clarke, whose novel *The Fountains of Paradise* (Gollancz, 1979) features the 22 300-mile orbital tower, revealed that these paths of thought had been traversed by others. In a letter to *Science* (Vol. 158, 17 November 1967, p. 946) Vladimir Lvov of the Novosti Press Agency, USSR, claimed priority on the orbital tower for a Russian pioneer of 1895, Konstantin Tsiolkovski; and attributed the orbital cable to the Leningrad engineer Yu. Artsutanov, who published an account of the idea in *Komsomolskaya Pravda* (31 July 1960). The lunar cable seems to have been originated by S. W. Golomb in 1962 (*Astronautics*, Vol. 7 (8), 1962, p. 26).

Arthur C. Clarke tells me that 'most of the work in this field is now being done by Jerome Pearson at the USAF Wright-Patterson base, Ohio. His first paper was entitled "The Orbital Tower; a spacecraft launcher using the Earth's rotational energy" (*Acta Astronautica*, Vol. 2, 1975, p. 785). He thought he'd invented the idea, a computer search having failed to turn up even the Isaacs paper!'

And there, pending a re-examination of the notebooks of Leonardo da Vinci, the matter rests at present. My only consolation is that DREADCO was one of the many independent originators of this mighty challenge to funicular technology.

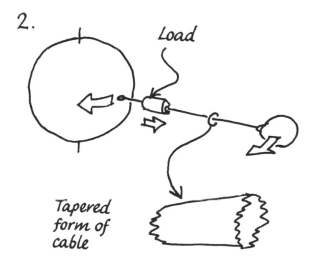

1.

1. The basic synchronous satellite

A small satellite orbits in a period P given by $P^2 = 4\pi^2 a^3 / GM$, where

    a is the radius of its orbit,

    G is the universal gravitational constant (6.67×10^{-11} $\mathrm{m^3 kg^{-1} s^{-2}}$), and

    M is the mass of its primary (5.97×10^{24} kg for the Earth).

    Since the Earth spins once a day (86 400 s), putting $P = 86\,400$ in this equation gives $a = 42\,230$ km or (since $r_0 = 6370$ km) $h = 35\,860$ km $= 22\,300$ miles, as the 'synchronous altitude' at which a satellite will remain stationary above a point on the Equator

2.

Load

Tapered form of cable

2. The tethered synchronous satellite

Each section of cable must be strong enough to support the whole mass below it. This implies a tapered form, which Isaacs et al. (loc. cit) show should be approximately exponential. The cross-section A_0 at surface radius r_0 is related to that at synchronous radius A_a by $A_0 \simeq A_0 \exp(\rho r_0 g / Y)$, where

    ρ is the density of the cable,

    g is the gravitational acceleration at the surface of the Earth, and

    Y is the yield stress per unit cross-section of the material of the cable

3.

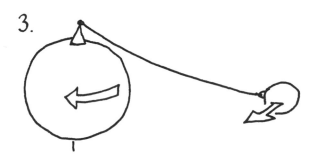

3. The tethered satellite on a pivot

This system enables satellites of any orbital period (e.g. the Moon) to be tethered. The forces on the pivot may, however, be severe

Salty spectra

When an electron is accelerated or made to change direction, it emits electromagnetic radiation. So, says Daedalus, an appropriate direct current flowing in a wiggly wire should emit light. Its frequency would equal the number of wiggles each electron traversed per second, and in this simple set-up would be a bit low for visibility. Even if the electrons moved at nearly the speed of light — as perhaps they might do in a long wiggly radio-valve under high voltage — they would only emit radiation of the same wavelength as the wiggles! To get visible light by this ingenious method one would have to find a tube whose wiggles were much shorter than a wavelength of light. Daedalus reckons that crystalline salt is a good candidate. It has positive sodium ions and negative chlorine ions packed alternately in a 0.28-nanometre spacing, so an electron beam racing over a salt surface should be violently waggled by the alternation of charges. Daedalus calculates that a salt-filled radio-valve should emit visible light at only 0.05 V of applied potential.

Efficiency would be very low, however, for the electrons would spend very little of their journey skimming the crystal faces. So Daedalus sought some hollow ionic-lattice material with tortuous internal passages through which the electrons would be forced to wander. The zeolites are the obvious choice — the materials from which ion-exchange media and molecular sieves are made. Their convoluted internal structure would make an ideal 'wiggly maze' for electrons, with wiggles about 2 nm in wavelength. So Daedalus is sticking electrodes onto chunks of zeolite, sealing them into glass envelopes, and pumping them down to high vacuum. The result will be a molecularly tortuous radio-valve. The zeolite will act as a sort of charge-bearing maze for electrons, which will zig-zag madly from cathode to anode, emitting furiously as they go. About 3 V will give visible radiation, but this ingenious light-source is infinitely tunable. It can give anything from deep red to far ultraviolet depending on the voltage applied, which governs the speed of the electrons and hence the frequency at which they traverse the molecular wiggles of the zeolite.

The ability of this new 'Zeolamp' to rainbow through the spectrum as fast as its voltage can be changed, will give it important uses in spectroscopy and in domestic illumination. On normal a.c. mains it will seem to have a steady colour: its sweep through the spectrum and back 50 times a second will be too fast to resolve. But the resulting time-averaged hue will be easily variable by voltage or waveform biasing, altering the instantaneous contributions of the different parts of the spectrum. Theatre and show-biz circles generally will welcome the Zeolamp for its complete command of colour. In particular Daedalus sees it as a weapon in the live concert's fight against hi-fi. For a Zeolamp fed with the amplified audio-signal of a musical instrument would act as a marvellous colour-sweeping strobe-lamp. It would illuminate the moving string or vibrating surface with a colour that changed in phase with the movement, making violin or drum or cymbal a wonderful solid rainbow of colour, iridescing subtly with every changing chord.

(*New Scientist*, 25 July 1974)

From Daedalus's notebook

An electron of mass m and charge e, accelerated through a potential E, will acquire a velocity v given by:

$$Ee = \tfrac{1}{2}mv^2$$

Imagine it traversing a wiggly lattice of lattice-spacing l. If we want it to wiggle at frequency ν in the process, it must traverse ν lattice-spacings per second, i.e. it must have velocity $v = \nu l$. The potential needed to do this will be given by:

$$\begin{aligned} E &= \tfrac{1}{2}mv^2/e \\ &= m\nu^2 l^2/2e \\ &= k\nu^2 l^2 \end{aligned}$$

With $m = 9.11 \times 10^{-31}$ kg and $e = 1.60 \times 10^{-19}$ C, we find $k = 2.8 \times 10^{-12}$ kg C^{-1}.

So, for example, to generate yellow light of $\nu = 500$ THz by firing electrons across a crystal of salt with lattice-spacing $l = 0.28$ nm, we need $E = 2.8 \times 10^{-12} \times (500 \times 10^{12})^2 \times (0.28 \times 10^{-9})^2 = 0.054$ V; a bit low for copious electron-emission from common cathode materials.

The zeolites are a much better bet. They are translucent, and their connected-chamber structure presents electrons with a volume rather than a surface to wander in:

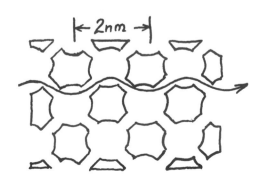

And with l now about 2 nm, the voltages come out much more conveniently. Red 400 THz light needs 1.8 V, yellow 500 THz light needs 2.8 V, blue-green 600 THz light needs 4 V, and violet 750 THz light needs 6.3 V. So it will be easy to range far into the ultraviolet by jacking up the voltage, to several kilovolts if necessary, but hard to get into the infrared without the voltages getting too low for easy electron-emission. The Zeolamp is essentially a tunable UV/visible source.

Daedalus comments

This invention isn't as brilliantly novel as I thought at the time. I later found that much the same principle is used in the Smith–Purcell source for far-IR radiation (S. J. Smith and E. M. Purcell, *Physical Review*, Vol. 92, 1953, p. 1069). In this device an electron-beam is fired over the surface of a diffraction-grating ruled with many lines per millimetre, and oscillates to follow the grating surface. But a diffraction-grating has far coarser wiggles than a zeolite lattice, and the Smith–Purcell source has been advocated only for generating very long-wavelength IR radiation. It is also a two-dimensional affair, whereas the Zeolamp should generate light throughout its whole volume.

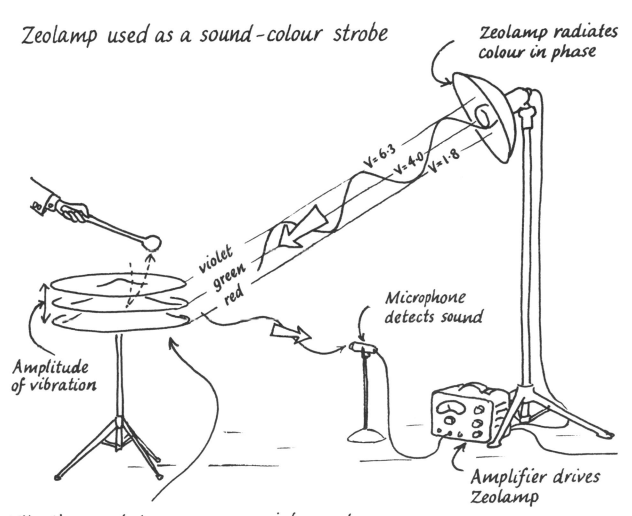

Zeolamp used as a sound-colour strobe

Zeolamp radiates colour in phase

V = 6.3 V = 4.0 V = 1.8

violet green red

Microphone detects sound

Amplitude of vibration

Amplifier drives Zeolamp

Vibrating cymbal appears as a rainbow-cake, violet at the upper limit of amplitude and red at the lower limit. In practice more complex harmonics would give a richer visual effect.

Semicircadian rhythms

We all carry inside us some sort of 'internal clock' which governs our rhythm of waking and sleeping. This inbuilt 'circadian rhythm' is normally locked to the day/night cycle, but is not absolutely steady. It can be jarred by jet-lag; in steady light or darkness it runs unguided at only approximately one cycle per 24 hours. Daedalus is exploring the idea that it is basically obtained by dividing-down the heart-beat, as a clock divides down a pendulum-swing. Division by 2 is the easiest form of dividing, and 17 successive divisions by 2 would generate the circadian rhythm from the pulse-rate quite plausibly. Daedalus claims that development and ageing, the natural periods of life, are further programmed from the circadian rhythm on exactly the same principle. Thus, starting from birth, 12 successive binary divisions of the circadian rhythm initiate puberty, and a further 2 initiate the menopause in women; a further binary division is generally fatal. The interest of all this is that many clocks and oscillator-dividers can be locked onto the first harmonic of the drive-frequency, thus running exactly twice as fast. Alternatively, by slipping every other cycle they can lock onto half that frequency and run twice as slow. So while the human circadian rhythm strongly resists attempts to make it go a little bit fast or a little bit slow, more drastic twofold alterations may be quite easy.

Accordingly Daedalus plans to revolutionize education and human development by modifying its timing. He is designing schools and homes lit entirely by artificial light in which speeded-up days of 6 hours light and 6 dark, or extended days of 24 hours light and 24 dark, can be maintained. The childrens' circadian rhythms should soon lock onto these and operate happily at the first-harmonic or first-submultiple frequencies. The slow rate would be ideal for periods like the language-mastery span around 3 to 5, or the 'intelligence explosion' of 14 to 15 or so, where so much more could be achieved if there were more time. Conversely, the rapid rate would quickly get over the awkward period of early adolescence, or that period of maximum exasperation around 18 months in babies.

Other periods of life could benefit too. The 'mid-life crisis' of so many executives, who realize at last how long and how thoroughly they have been conned by the system, could be mercifully compressed; and the fruitful period of later creativity that many men enjoy in the tranquillity of their riper years, could be extended. But Daedalus suspects that running a whole lifetime at half-speed and living to be 140, or setting the clock into reverse and regressing backwards to your first childhood, would somehow not work.

(*New Scientist,* 14 August 1975)

From Daedalus's notebook

Most digital clocks, computer-timers, etc., count by the repetitive binary division of the signal from a primary oscillator, using a chain of flip-flop circuits. Now neurones, as digital circuit elements, could easily be assembled into a train of flip-flops. So if the brain wanted highly accurate timing, it could work this trick too. Does it?

In the animal kingdom, the supreme exponent of timing must be the 17-year cicada (harvest-fly) which spends 16 years underground in larval form, and emerges to mate in its exact 17th year. It must surely use digital timing. And in man, the biorhythm theorists claim that three distinct cycles are initiated at birth, and run accurately for the rest of life: an intellectual cycle of 33 days, an emotional one of 28 days and a physical one of 20 days. Again, accuracy of this order could only be obtained by digital timing.

So let's look for possible cycles. The brain's alpha-rhythm at some 10–11 Hz is a good candidate for the title of master oscillator. Three successive divisions by 2 give about 1.3 Hz or say 80 beats per minute as a digitally derived heartbeat. The resting heart-rate is about 70 beats per minute so a typical fairly active day could well keep the heart to 80 beats per minute or 1.3 Hz on average. To get a circadian rhythm from this we need a further 17 divisions by two: $1.3/(2^{17}) = 10^{-5}$ Hz or 1 cycle per 28 hours. This isn't a bad estimate of the free-running circadian rhythm. Record-breaking pothole-squatters always underestimate the time they have spent underground, and their free-running rhythm must be of this order. In normal conditions the oscillator must be 'slipped' a bit daily, to stay in phase with the 24-hour day.

We can get the 33-day intellectual biorhythm to within a day by dividing the true day-period by 2^5, and the 20-day physical cycle to similar accuracy by dividing the 28-hour circadian rhythm by 2^4. The 28-day emotional cycle may be e.g. a submultiple of an intermodulation sum ($\frac{1}{2}(33 + 20) = 26.5$?). It's all a bit sloppy, as you'd expect. Biological electronics is never designed for needlessly high accuracy, and must tolerate a lot of missed or extra beats and environmental resetting. The biorhythm enthusiasts are probably overplaying their hand.

Daedalus comments

Biorhythm theory seems to be taken seriously by the US computer company, Control Data Corporation (*New Scientist,* 1 January 1981, p. 38). But for a more jaundiced viewpoint, see J.W. Shaffer in *Archives of General Psychiatry* (Vol. 35 (1), 1978, p.41) and *New Scientist* (20 March 1980, p. 926).

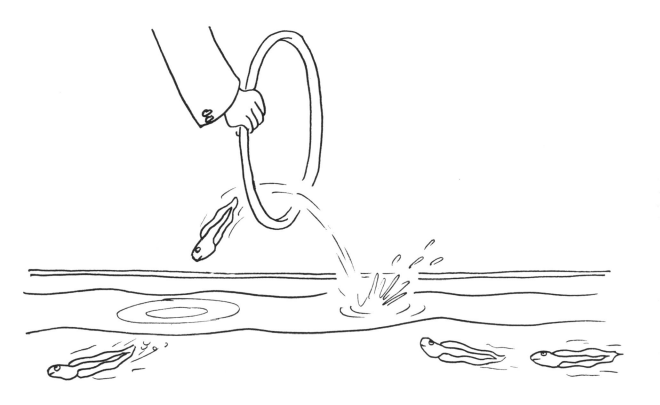

The DREADCO Biorhythms Group has had some success in delaying the development of various animals, and thus extending their youthful learning-period, which in most species is cruelly cut short by rapid maturation. After seven years of training, these performing tadpoles can swim in formation and jump through a hoop

Magnetic fur

Architects seeking to improve the thermal insulation of their buildings should take a leaf out of Nature's book. Instead of putting insulation on the inside or in wall cavities, it should be stuck on the outside. Heat would then be retained in the brickwork, whose massive thermal inertia would level out day/night fluctuations in temperature and thus save even more energy. Ideally, as in nature, the perfect insulant should also be water-repelling, sound-deadening, decorative, and protective. The problems of making a furry house are, however, considerable. Sticking glass wool all over it is very crude, and electrostatically flock-spraying a pile on it would also be tricky. DREADCO biologists have been seeking single-filament plants which might be grown onto a wall as a sort of vegetable fleece, but fear that like ivy, they will slowly attack the brickwork.

The answer, says Daedalus, is a novel magnetic paint. It is a suspension of iron filings in a solvent-thinned polymer lacquer. The paint is coated onto a surface, and a powerful magnet is passed over it while it is still wet. Each filing is pulled out of the surface towards the magnet, drawing a fine filament of lacquer behind it. This sets almost instantaneously by rapid evaporation, leaving a long, fine polymer pile on the new paint. The resulting surface — which can, of course, have any colour — will consist of fine droopy fibres, readily shedding water in thatch fashion, and absorbent to sound and wind-noise. In addition to its splendid insulative and protective properties, it will be very beautiful. A green furry house would blend ideally into a rural background, and the wind rippling over its surface would be a constant visual charm. Even in towns, furry tower-blocks would have a softness much kinder to the eye than the current monstrosities. Tabby, tortoiseshell, zebra and other fleecy designs could easily be created from appropriate base patterns, and would give a new visual and tactile appeal to all architecture. Daedalus also intends to exploit his invention in more personal areas. DREADCO's anti-baldness lacquer (just brush on and move magnet in desired pattern over scalp. Anchors as firmly as your own hair!) should command a ready sale. It might also solve the sartorial problems of arctic explorers and first-hand investigators of gorilla sociology.

But Daedalus feels that these purely passive applications in thermal insulation fail to exploit his new product completely. The iron filing at the top of each fibre could be moved by a magnetic field, which opens up a host of entirely new possibilities. Thus magnetic fur could be made into the first controllable thermal insulator. A furry house, or even a furry shirt, could be controlled by a pattern of fine electric wires woven into it. In hot weather, a thermostat could turn on the current to lay the pile flat; in could weather it could fluff up for heat-retention. Thus the house could take active advantage of the weather, and the shirt-wearer would enjoy the automatic thermoregulation available to cats and other furry creatures. If the fibres also extended inside the shirt, local a.c. pulses from a discreet wrist panel could vibrate selected areas of fibres, enabling the wearer to scratch himself by remote control in places otherwise accessible only with indignity. Again, as a novel bathing-dress the shirt could 'swim' the wearer through the water bacterium-like, by rapid rowing of its magnetic 'cilia'; and afterwards the same action would enable him to shake himself clean and dry in true doggy fashion.

Even more interesting, fine magnetic fibres could vibrate in sympathy with an a.c. field even up to the very highest frequencies of the audio range. A loudspeaking shirt, taking its cue from microphone and microchip amplifier in the collar, would appeal to orators; but the principle would find wider application in loudspeaking wallpaper. Daedalus envisages a wall with a grid of wires in it carrying the signal current and driving the magnetic fur wallpaper in sympathy. This would couple to the air by viscous drag (a very efficient mechanism for fine fibres) giving completely distributed, non-directional hi-fi. If different sections of the wall carried different signals, true multichannel polyphonic sound would be possible. Other inventions suggested by this marvellous new development include a magnetic-bristle toothbrush which does the scrubbing for you, and a peristaltic carpet. This ingenious object again has a magnetic-fur pile and woven-in control wires. Its signal generator creates a continuous stream of ripples traversing the pile, carrying fluff, dust and light litter towards a small dustpan where the ripples converge. Much housework will be eliminated. But cats will hate it.

(*New Scientist*, 6 and 13 June 1974)

Raising a magnetic pile

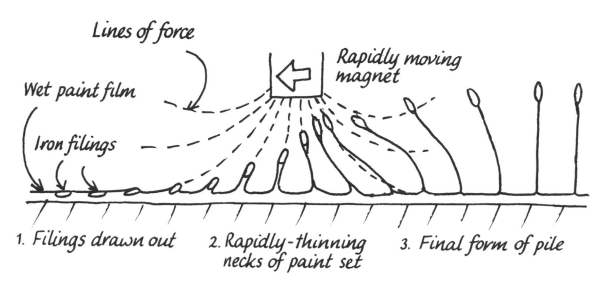

Lines of force

Wet paint film

Iron filings

Rapidly moving magnet

1. Filings drawn out   2. Rapidly-thinning necks of paint set   3. Final form of pile

PERISTALTIC CARPET

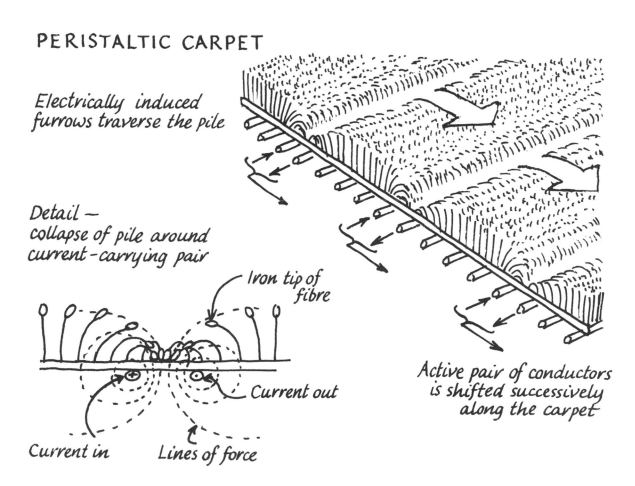

Electrically induced furrows traverse the pile

Detail — collapse of pile around current-carrying pair

Iron tip of fibre

Current out

Current in   Lines of force

Active pair of conductors is shifted successively along the carpet

Audible vertigo

Our ears are dual-purpose organs. They deal simultaneously with hearing (in the cochlea) and with balancing (in the semicircular canal labyrinth). Received sound must leak from the cochlea and perturb the connected semicircular canals — why doesn't a shout throw us off balance? Ah, says Daedalus, the rapid oscillations of sound average to zero in the balancing labyrinth, and cause no net movement of its fluid. But, he muses, suppose you presented the ear with a highly asymmetric spiked wave-form, with sudden pulses of high pressure interspersed with much longer periods of steady lower pressure? Daedalus reckons that this sound wouldn't average to zero in the semicircular canals, but would be 'rectified' to steady flow. The visco-elastic ear-fluid would be unable to follow the sudden pulses one way, but would flow with the steady pressure of the opposite part of the cycle.

At first Daedalus was unhappy with this deduction. Spiky waveforms are generated in audio laboratories all the time without causing vertigo and collapse in the scientific staff. But he then recalled that amplifiers, speakers and indeed ears, all introduce strong phase-shifts into the signals they handle. They retain the frequency-spectrum which determines the audible quality of the signal, but distort the actual waveform out of all recognition. So DREADCO engineers are building special oscillators to generate subtle sound-shapes with exactly contrary phase-shifts, chosen in such a way that, after suffering the various phase-vicissitudes of their journey, they will arrive at the ear-labyrinth as true spiked waveforms. They will set up flow in the semicircular canals; the hearer, fed with false balancing information, will feel giddy and fall over. This terrifying non-violent weapon would be ideal for quelling crowds, breaking up columns of soldiers, and indeed for making anybody stop what they are doing and hang on to the nearest solid object for support. Its tactical value would be somewhat reduced by the fact that it would affect both sides equally. Lower intensities of 'vertigo-sound' would cause a controllable queasiness which might be welcomed by avant-garde composers still trying to shock the bourgeoisie. Even subtler, *ultrasonic* vertigo-sound would do its work in apparent silence. Aversion-therapy psychiatrists would welcome a remote-control method of making alcoholics, drug-addicts, exposurists, etc., feel instantly uncomfortable the moment they indulge their weakness. And Daedalus himself entertains dark fantasies of sneaking strong vertigosound into the sound-track of *Coronation Street* and bringing the whole nation literally to its knees!

But the technology of vertigosound could be beneficial too. There is no reason why the balance-information it feeds to the inner ears has to be false. An obvious useful application, therefore, is to use the method in a balance-amplifier, a sort of deaf-aid, to feed the semicircular canals with stronger and more exact data than they can generate for themselves. The elderly, and those with defective inner ears, would benefit greatly. Furthermore, says Daedalus, it should be possible not merely to remedy the failings of Nature, but to improve on her dramatically. For while as sound-detectors our two ears interact constructively to give us a sonic stereo-image that one ear alone cannot provide, our two semicircular canal labyrinths merely act in parallel. They detect the rate of rotation of the head about an axis but do not locate the centre of spin. But by feeding them with slightly different balance-data, which summed to a true average but exaggerated the differences, they should be able to register not merely angular accelerations in the three directions but also instantaneous centres of rotation. So Daedalus is designing a 'stereo-hat' carrying modern sensitive accelerometers. It will measure angular velocities and accelerations about all three axes, locate the centres of rotation, and send its findings to left and right earphones encoded as appropriate asymmetric ultrasonic waveforms. The phones will not be totally enclosed, so the wearer will be able to hear normally, and he will not be disturbed by the extra ultrasonic signals. But he will be receiving new and detailed information about his position and movement. The ease with which we learn to ride a bicycle shows how rapidly new balancing skills become automatic, so wearers of the stereo-hat will rapidly come to incorporate its accurate findings into their reflexes.

Thus a new generation of dancers, trapeze-artists, steeplejacks and unicyclists will be born. Old age pensioners will grace the high wire; sport will be transformed; a new delight in spatial awareness will enrich all our lives! More intriguing still, special plug-ins will enable the stereo-hat to relay information from other instruments. Airline pilots could be made directly, stereoscopically, aware of the attitude of their machine even in fog or darkness. Subjectively, they would *become* the aeroplane, and would guide it with instinctive and unerring sureness. Astronauts too, who can suffer from terrible inner-ear disorientation, could be given secure subjective inertial navigation. Even sufferers from seasickness could be effortlessly rescued by feeding their stereo-hats from a stable artificial horizon. Conversely, in cases of accidental poisoning a contradictory signal-pattern from the stereo-hat could promote instant strong nausea, ejecting the deadly stomach-contents without delay.

(*New Scientist*, 16 and 23 November 1978)

Vertigosound could clear public buildings humanely of starlings, pigeons, etc., by making the birds overbalance from the narrow ledges

Bacterial plastic macs

According to the genetic engineers, bacteria given the right DNA will make almost any biochemical, especially expensive pharmaceuticals. But to get the medicine out of the bug and into the patient, culturing, extraction, purification and storage of the product will be needed, followed by subsequent injection. Daedalus plans to short-cut the whole process by inventing the bacterial overcoat. You can't just inject bacteria into people because the body defences attack them, and they fight back by multiplying furiously. But, says Daedalus, imagine a bacterium encapsulated in a thin polymer membrane. Its foreign surface proteins, which trigger immunological attack, are safely screened. If the membrane is adequately permeable to water, nutrients, and metabolic products, the bacterium will be able to live happily enough. But it won't be able to grow big enough to divide into two because there's no room. Since bacteria seem essentially immortal, overcoated bacteria in a patient will circulate in his bloodstream pumping out medication for ever: or at any rate for months, until degradation or rupture of their envelopes lets them out one by one. Such individual invaders, unlike the mass-attack of a conventional infection, will of course be mopped up by the body's defences with no trouble at all.

So DREADCO biochemists are devising polymerization-catalysts which can be adsorbed on cell-walls, so as to form thin bacterial overcoats *in situ* from solutions of suitable monomers. Once they have the technique working, overcoated bacteria will soon be circulating in test patients, pumping out insulin, interferon, antibiotics or contraceptive steroids. One injection will bring months of automatic medication. Even pharmaceuticals far too unstable to survive traditional preparation and shelf-storage will be usable. In this connection Daedalus recalls that nicotine has to be smoked because it is unstable in the air. Only by rapid thermal release from its plant-precursor, and swift inhalation, does enough get into the lungs to affect its user. So internal overcoated bacteria, permanently exuding low levels of freshly synthesized nicotine, could bring stable satisfaction to the smoker, save him from tar-induced lung cancer, and free the rest of us from his smell.

(*New Scientist*, 5 February 1981)

From Daedalus's notebook

Can bacteria survive in polymer overcoats? The answer seems to be yes, judging by the work of S.J. Updike, D.R. Harris and E. Shrago (*Nature*, Vol. 224, 1969, p.1122). These workers put *Tetrahymena pyriformis* and *Escherichia coli* into solutions of acrylamide, and then set the solutions to solid polyacrylamide gel by fast photopolymerization. The *T. pyriformis* seemed to be struggling in their polymer cages, poor things. The organisms stayed alive for up to 5 days, although the gel can't have been all that nutritious. In thin permeable jackets their chances should be much better.

Would a polymer overcoat be adequately permeable to water, oxygen, metabolites, etc? Suppose we have an *E. coli* of (say) 3 μm diameter (1.5×10^{-6} m radius). Its surface area is $A = 4\pi r^2 = 2.8 \times 10^{-11}$ m^2; its volume $V = 4\pi r^3/3 = 1.4 \times 10^{-17}$ m^3. A typical LD polythene as a 25 μ film has a permeability to water-vapour under 75% relative humidity differential of 2×10^{-7} kg m^{-2}s^{-1}; so a film 1 μm thick would pass $k = 25 \times 2 \times 10^{-7}$ kg m^{-2}s^{-1}/(1000 kg m^{-3}) $= 5 \times 10^{-9}$ m^3 of water per m^2 per second. So an *E. coli* organism in a 1 μm LD polythene 'raincoat' could exchange its own volume of water through the raincoat in time $t = V/kA = 1.4 \times 10^{-17}/(5 \times 10^{-9} \times 2.8 \times 10^{-11}) = 100$ seconds. The flow-rate will be lower for bigger molecules than water, but there's clearly no fundamental problem of permeability. We may have to stick, e.g., amine sidechains on the polymer to increase its permeability to insulin or any other proteins that have to get through. This would also aid the body to attack and dispose of it when it finally degrades and splits and lets its occupant out.

What sort of products could raincoated guest bacteria most usefully make? Drugs for permanent therapy effective on the milligram level (higher outputs might need inconveniently many bacteria). So: insulin, contraceptive steroids, modicate, valium (for chronic neurotics). Vitamins too — the body can't make them itself, so it would be highly convenient to have them permanently supplied by guest bacteria. Even social drugs — caffeine, nicotine, cannabinol, perhaps even heroin or cocaine — could as stable presences in the body save many people from an awful lot of bother, much of it expensive and antisocial. But alcohol, which must be taken in tens of grams to have any worthwhile effect, might just require too many organisms.

One neat trick — feed some bacterium the DNA for synthesizing penicillin. This antibiotic only attacks bacteria while they are dividing. So as long as the bug stays monastically inside its 'plastic mac' it will come to no harm. But when at the end of the treatment the weakening plastic starts to split and lets it out, it will begin to divide and instantly commit suicide.

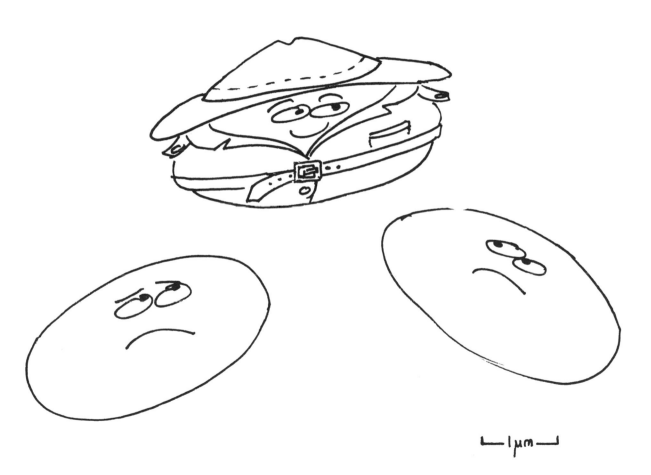

Dynachair

While attending an interminable scientific lecture recently, and enduring posterior anguish from the trendily designed chairs of the lecture-hall, Daedalus began elaborating a dynamic theory of comfort which may revolutionize the world of furniture design. He reasoned that even the best designs of chair become irksome after a time because of sensory deprivation. Nerves and muscles look for variety in their life and no static loading, however equable, can be permanently pleasant. Attempts to provide this variety by squirming about on the chair never really succeed. So Daedalus is devising his dynamic chair, the seat, back, and sides of which consist of a number of independent elements all moving slowly and irregularly in a pattern optimized to the sensory needs of the human posterior. Even the texture of the seat material slowly changes as the many tiny fabric tubes of its construction, each covered with a different material and fed from a separate gas-supply, randomly inflate or collapse. At last dignity and long-term comfort need not conflict!

DREADCO hopes for brisk sales to conference-halls, airlines, schools, court-rooms, Houses of Parliament, etc., using the slogan '*Dynachair* — the chair that does the fidgeting for you!' The principle can easily be extended to other regions of the human anatomy. Soldiers, for obscure reasons, are often required to stand quite still for long periods of time: they should welcome DREADCO's wriggle-boots, whose independently varying heel height and angle of tilt give the feet ample rest and exercise to avoid sensory monotony. A similar writhing-attachment for underwear could eliminate at source much of the itching and constriction that such garments seem to generate during interviews, sermons, hair-cuts and other tense static situations. And the DREADCO dynamic bed, by providing constant slow sensory variety for the sleeper, should eliminate the endless slow tossing and turning which makes a speeded-up film of a person asleep look as if he is wrestling with an invisible boa-constrictor. Sleep of a hitherto unknown depth and relaxation will result.

(*New Scientist*, 16 December 1971)

Daedalus comments

Once again, it seems that DREADCO has pioneered a development that others have later been glad to copy. Various dynamic beds have since been patented or are under development (Hunkin, who brings together so many intriguing snippets in his *Observer Magazine* strip 'The Rudiments of Wisdom', mentions a few of them in his contribution of 4 May 1975). And Du Pont's *Elastomers Notebook* for December 1978 records this use of the Company's 'Hytrel' elastomeric polyester:

PULSATING SEAT reduces driver fatigue by promoting blood circulation in the buttocks and legs. Developed by H. Koch and Sons for U.S. military flight personnel, the PULSAIR™ seat cushion demonstrated its value at Brooks Air Force School of Aerospace Medicine. The device has recently been adapted to use by truckers. It consists of a thin bladder of Du Pont *Hytrel* polyester elastomer with a network of internal air chambers that are sequentially inflated for two seconds and then deflated for six seconds. This produces a scarcely perceptible wave motion that gently massages muscles and stimulates blood flow. It is timed to operate ten minutes out of every hour.

Even the pneumatic mechanism I proposed for the Dynachair is imitated in this American version!

But DREADCO, as always, remains one jump ahead. I have since decided that random sensory stimulation is not enough, and that the optimum dynamic chair must employ feedback principles. My current musings on this topic suggest sensors in the chair to detect the small local muscle tremors that give warning of incipient discomfort or sensory boredom in that part of the sitter's anatomy. The chair then generates movements in the area until it finds the change needed to make the tremors vanish. Thus it anticipates your sensory and muscular needs, and meets them without effort or even conscious attention on your part. It really *does* fidget for you!

The anti-greenhouse

So far, all attempts to exploit solar energy have concentrated on capturing the Sun's rays. To Daedalus, this approach is understandable but very naive. You can't generate power from heat alone. You need both heat *and* cold; then you can use the thermal flow from one to the other to drive an engine. Every heat engine not only needs a 'boiler' of some sort, but a 'condenser' as well.

Daedalus, recalling the basic thermodynamic fact that the efficiency of a heat engine depends on the difference in temperature between boiler and condenser, has been examining the available 'condensers' for the solar heat-engine. The best of all such condensers is the ultimate radiation-sink of the night sky. It is totally black and (apart from a little starlight) is effectively at absolute zero. So Daedalus's scheme is to build a condenser system with a big concave mirror under it so that at night it is optically completely surrounded by night sky. If it is kept in a transparent vacuum-jacket to prevent warming by contact with the air it should then cool right down towards absolute zero as its radiation is lost in space. It would then serve to liquefy air and thus act as the condenser for a liquid–air engine whose boiler would be at normal ambient Earth temperature. This scheme does away with the elaborate optics needed by the usual solar power schemes, for any big, inaccurate mirror will do to reflect the vast sky, and no focusing or tracking is required. In this system, the Earth itself is the 'boiler'. The Sun merely plays its customary role of keeping the Earth warm despite the steady loss of terrestrial radiant heat into space.

Daedalus has also invented the anti-greenhouse to enable the system to keep operating in the daytime. This enclosure, exactly opposite to the greenhouse, is constructed of a material like black polythene or cadmium telluride. These substances are opaque to the short visible wavelengths emitted by the Sun, but are transparent to the longer wavelengths radiated by objects cooling from ambient temperatures. So the anti-greenhouse won't let the sunlight in, but will let internal heat-radiation out; it will therefore cool steadily down. As the daytime sky is only bright by scattered sunlight the anti-greenhouse will receive no heat from this source either. So it will continue to lose heat to space night and day, and will generate cold and cryogenic power all the time.

(*New Scientist*, 3 March 1966)

Daedalus comments

The earliest primitive anti-greenhouse was, I suppose, the dewpond. This was simply a large depression in a mass of impermeable rock. On clear nights the rock radiated heat away to the cold sky; the air in contact with the rock cooled and deposited its humidity. My mirror scheme simply takes this idea to the limit. Just as an accurately parabolic concave mirror can optically 'wrap' the Sun around an object and thus heat it to red heat, so a big mirror can surround an object completely by night sky. The optics are very much simpler than in the solar case, and as a power generator the system could in principle be more efficient. For the ultimate Carnot efficiency of any thermal engine is $\eta = (T_b - T_c)/T_b$, where T_b is the boiler temperature and T_c the condenser temperature, both in degrees absolute. Clearly reducing T_c towards zero pushes η up towards 1 much faster than increasing T_b. To attain 50% efficiency with a conventional solar engine using the ambient environment at 300 K as a condenser requires $T_b = 600$ K. Using ambient as a boiler, 50% efficiency is attained at $T_c = 150$ K, halving the temperature difference to be maintained.

The enclosed anti-greenhouse is an obvious next development, using filters rather than mirrors to reject solar radiation while radiating maximally to space. By matching the filters carefully enough to the wavelengths of atmospheric transparency, daytime cooling can be attained too. And happily, since the original column appeared, the first anti-greenhouse on these lines has been experimentally tested. B. Bartoli and his co-workers at the University of Naples (*Applied Energy*, Vol. 3, 1978, p. 267) have devised filters rather superior to the cadmium telluride I advocated. They use polished aluminium covered by Tedlar film, which is transparent to visible light but black to the 8–13 μm transparent 'infrared window' of the atmosphere. Such surfaces therefore reject sunlight, which 'sees' the reflective metal, but lose heat by infrared radiation to space. Experimental panels have stayed 10 deg C below ambient both day and night, and further improvement seems possible. A combination of mirrors and filters should in principle be able to cool a specimen right down to the 3 K cosmic background, and I await developments with interest.

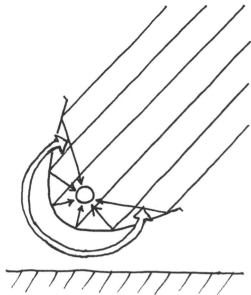

Extent to which a precise paraboloid can 'surround' a small object by hot sun.

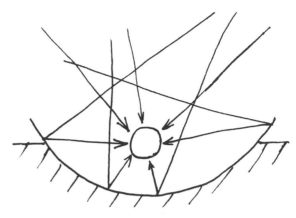

A very rough mirror can 'surround' a large object completely by cold sky, with no tracking needed.

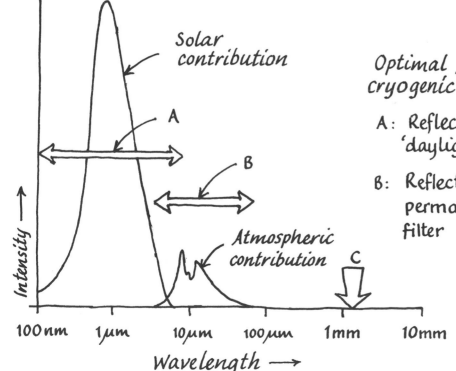

Radiation received from the sky

Solar contribution

A

B

Atmospheric contribution

C

Intensity →

100 nm   1 μm   10 μm   100 μm   1 mm   10 mm

Wavelength →

Optimal filters for cryogenic anti-greenhouse

A: Reflective region of 'daylight-only' filter

B: Reflective region of permanent (day & night) filter

C: Frequency of maximum emission at 3 K

Galvanized plants

All life ultimately depends on photosynthesis in green plants; so their wasteful use of light, with photochemical efficiencies of only a few per cent, is rather distressing. Daedalus points out how severely the growth of plants is limited by their sluggish internal flow of nutrients and water. If transpiration, that gentle flow of sap-water up from the roots to evaporate from the leaves, could be increased, the whole metabolic rate of plants could be speeded up dramatically. Daedalus aims to do this by electro-osmosis, the phenomenon whereby water is impelled through a porous medium by an applied electrical potential. He calculates that a vertical potential gradient of a few thousand volts per metre could easily double transpiration rates. So a few hundred volts at the top of a blade of grass, a few kV on a wheat-stalk, or a hundred kV on a tall pine tree, could send them all shooting upwards. Furthermore, the high resistance of woody tissue means that the resulting leakage currents will be very low; accordingly only a few watts of power will be needed to maintain these high voltages. So DREADCO botanists are fixing wind-powered high-voltage generators ('Windhurst machines') on the tops of conifers, and stringing dense horizontal webs of electrified wire across wheat-fields (which should discourage the birds, too). Thus will agriculture be revolutionized. With wheat shooting up in weeks, trees reaching full size in a year, and beans racing dizzyingly around their electrified poles, a new era of vegetable plenty will open up for us all.

Electric plants will be easy to control as well. Cutting the voltage will slow their growth; reversing it will arrest sap-flow and send the plant into 'suspended vegetation'; still higher reverse voltage will send the sap the wrong way and kill it. So clearance of overgrown land will only need a bit of wire netting to throw over the mass, and a few croc-clips to apply the power. All those nasty persistent chemicals will be unnecessary. Similarly, privet hedges could be neatly trimmed, and even formed into high-precision topiary, by an electrified former placed over the bush, to arrest each shoot as soon as it reached the exact position defined by the metal former. And the DREADCO electric grass-restrictor, a sheet of perforated zinc suspended an inch or so above the ground, will give the ultimate in lawn perfection. Every blade of grass will reach exactly the same length!

(*New Scientist*, 15 January 1981)

Electro-arrestive topiary. A wire former held at a retarding potential will arrest the growth of any shoot which touches it. The bush therefore replicates exactly the chosen shape

From Daedalus's notebook

Transpiration in plants has to pump sap upwards against gravity. So the pressure already developed by the plant must be at least equal and opposite to that resulting from the sap's barometric height: 1 atmosphere per 10 metres or some $10^4 \mathrm{N\,m^{-2}}$ per metre. To augment sap-flow significantly, electro-osmosis must contribute an additional upward pressure-gradient of at least the same order of magnitude. What voltage will be required?

The pressure p, in $\mathrm{N\,m^{-2}}$, exerted by a potential V volts across a liquid of relative permittivity ϵ in a porous medium of capillary radius r metres, with which it has an interfacial zeta-potential of ζ volts, is:

$$p = 8\ V\zeta\epsilon\epsilon_0/r^2$$

where ϵ_0 is the permittivity of free space, $8.85 \times 10^{-12}\,\mathrm{F\,m^{-1}}$. So a plant with xylem-ducts $20\,\mu\mathrm{m}$ in diameter (i.e. $r = 10^{-5}\mathrm{m}$ radius), containing aqueous sap of $\epsilon = 81$ at a (typical) zeta-potential of $\zeta = 0.05$ V will experience $10^4\mathrm{N\,m^{-2}}$ of electro-osmotic pressure

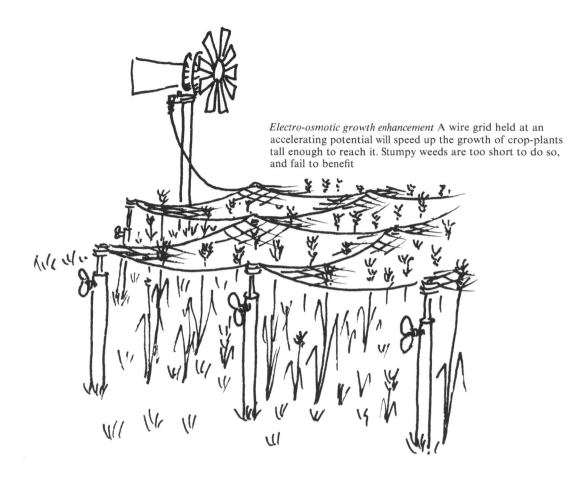

Electro-osmotic growth enhancement A wire grid held at an accelerating potential will speed up the growth of crop-plants tall enough to reach it. Stumpy weeds are too short to do so, and fail to benefit

on applying a voltage

$$V = pr^2/8\zeta\epsilon\epsilon_0$$
$$= 10^4 \times 10^{-10}/(8 \times 0.05 \times$$
$$81 \times 8.85 \times 10^{-12})$$
$$= 3500\,V$$

So about 3.5 kV must be applied for every vertical metre of plant stem to enhance transpiration by a useful amount. What power will be needed to maintain 3.5 kV m^{-1} in woody tissue?

The bulk resistivity of dry wood is around $10^8 - 10^{11}\,\Omega$ m, but living wood should have a lower resistivity: I'll guess about $10^6\,\Omega$ m. So a tree 5 cm in radius will have a resistance $R = 10^6/(\pi \times 0.05^2) = 1.3 \times 10^8$ ohms per metre of its height. In a field $E = 3.5$ kV m^{-1} it will dissipate power $P = E^2/R = 0.1$ W m^{-1}; a total of 1 W for a 10-m tree, at a current of $E/R = 30\,\mu$A. Smaller plants (shrubs, wheat-stalks, etc.) will clearly draw quite trivial amounts of power — thus a wheat-stalk of 1 mm^2 cross-section should dissipate about 13 μW per metre at a current of some 4 nA. All very promising!

Daedalus comments

Shortly after this item, based on the above calculation, appeared in *New Scientist*, the magazine received a reader's letter (12 February 1981, p. 456). It drew attention to two papers by V.H. Blackman in the *Journal of Agricultural Science* for 1924 (Vol. 14, pp. 240 and 268), describing the beneficial effects of high electrical potentials on the growth of cereals like oats and barley. All the conditions I had deduced were there — a web of wire above the plants, direct (rather than alternating) potentials, voltages of several tens of kilovolts, and currents of the order of nanoamps per individual cereal plant! Growth-rates increased by about 20% on average under these conditions.

My mortification at finding myself almost 60 years behind the times is mitigated by the fact that I seem to have provided a sound, quantitative theory for what was hitherto a mysterious effect. Since then (*New Scientist*, 12 February 1981, p. 406; 19 March 1981, p. 741) other experiments have been reported on plant-electrification. Maybe the field is about to take off again . . .

What goes up must come down

Daedalus notes with interest the recent recantation of Colonel Seifert, the high-rise architect who now sees such housing as socially evil. But what, short of adopting the desperate American solution of dynamiting the monstrosities, does he propose to do about it? Daedalus's own solution is very neat: don't knock them down, knock them over! Maybe prophetically, Colonel Seifert and his kind tended to set their housing blocks in gaunt, featureless parkland with ample space for each felled block to land on. In this way, they would be instantly converted into rows of dwellings just like the friendly terraces they were intended to replace. Of course you can't just fell a block like a chimney, even with a huge air-mattress laid out for it to land on. And the appealing dodge of filling it with helium to neutral buoyancy, cutting it loose and then manoeuvring it into horizontal place like a giant zeppelin, unfortunately fails even with the lightest constructions. So Daedalus is devising a system of retro-rockets and braking-parachutes by which, when the ground pillars are severed, the building tilts slowly and settles equably to earth in a flurry of smoke and billowing silk.

The refitting of a horizontal block poses intriguing problems. In a square stair-well, half the flights remain usable with treads and risers interchanged; the other half go sideways and must be realigned. The lifts become horizontal indoor railways, though with rather fearsome portcullis-type doors. Ceilings and floors become walls and vice versa, though luckily the modular and isotropic uniformity of modern design minimizes their differences. But baths and cupboards will exchange functions, and a lavatory is best re-used as a wall-mounted folded-horn hi-fi speaker. Much rerouting of plumbing will be needed, as well as the rectification of doors from their new role as vast letter-boxes and trapdoors. But the challenging life-style demanded by the new terraces should weld a genial cameraderie between the tenants and solve, at a stroke, the malaises of high-rise living.

(*New Scientist*, 5 December 1974)

Regarded as a structural material, snow is amazing stuff. The water-vapour in cold air condenses directly to crystals of snow without going through intermediate water; and the resulting feathery product aggregates to one of the lightest, most thermally insulating, and most sound-absorbing substances known. What a pity it melts at useful working temperatures! DREADCO chemists are trying to make artificial snows from other substances whose vapours condense directly to crystal. They are spraying iodine, naphthalene or ammonium chloride as hot vapours into a big controlled 'weather chamber' to produce falling 'snows' of these materials. But while naphthalene snow discourages moths, and iodine snow is a pretty purple, as structural materials they can only be temporary; for like snow itself they slowly sublime back to the vapour in a few weeks. What is needed is a snow that can be set. So the DREADCO team is seeking some suitable monomer which can not only condense from the vapour directly to the solid, but which can be polymerized to a non-evaporating plastic by ultra-violet light. Once condensed as a 'snow', it could then be made permanent by UV irradiation, when its molecules would link irreversibly into stable polymer.

This technology will revolutionize the construction industry. Instead of struggling with bricks, mortar, scaffolding, and Irishmen, the contractor will simply erect a big airtight tent on the site. Inside, the DREADCO snow-apparatus will create 'weather conditions' in which a steady stream of monomer vapour will condense to a continuous snowfall. It will fall onto a pattern of hollow inflatable room-moulds, covering them igloo-fashion with light, fluffy monomer snow. As each room is completed, it will be set solid by UV lamps, and the mould will be deflated, removed, and used to form the next room or storey of the structure. The finished polymer-snow house (or poligloo) should be quite rainproof. The water-repellent polymer surface, like those rainproofed fabrics, should exclude the heaviest downpour from its tiny pores while remaining slightly permeable to air. So poligloos will never be damp or stuffy, but always splendidly soundproofed and insulated. Indeed, claims Daedalus, a ventilation-system which draws air in through the walls would entrain any escaping heat and return it to the poligloo, giving total heat conservation and no fuel bills!

(*New Scientist*, 9 March 1978)

Daedalus comments

Shortly after this item appeared, Mr J. A. Faucher wrote to *New Scientist* with the news that the Union Carbide Corporation has had a vapour-deposited polymer ('Parylene', a poly-*p*-xylylene) on the market for over 10 years. He also helpfully enclosed technical literature which showed that Parylene, far from forming a snow, actually condenses to a solid remarkable for its freedom from holes. A nice try, Union Carbide. Keep at it!

ACACIA DRIVE·

Picfix Paper

The whole Prestel-Ceefax-Oracle business of relaying information onto TV screens is, in Daedalus's opinion, half-baked. You can't relax and study a TV screen, nor file it for future reference, nor compare it with yesterday's. Not until somebody invents a way of stripping a permanent image off a TV screen will the technology take off. So Daedalus has done just that. He points out that if you took a liquid-crystal display of the watch or calculator type, and cooled it down till the liquid crystal froze, the image it was displaying would freeze too. Furthermore, the image on a TV screen is created by a charged electron-beam and is therefore an image in voltage as well as light. So it would drive a liquid-crystal display laid over it. So DREADCO's 'Picfix Paper' (Regd) is just a thin layer of high-melting liquid crystal sandwiched between conducting and polarizing polymer sheets. Hold it against a TV screen, activate a warm-air or dielectric pulse heater, and the TV image is copied onto the paper! A moment later the liquid crystal has set solid again, and you have a stable copy.

Not only will this ingenious invention revolutionize the whole Prestel malarkey; it will break the universal dominion of paper. Newspapers, for example, could abandon printing and their vast wasteful appetite for paper and the woodpulp it is made from, and merely feed their page-images daily into the TV net. Readers could capture whatever they wanted and store it as permanently or as briefly as they liked. And when they'd finished with an image, they could re-use the Picfix to capture a new one. Picfix is perfectly erasable by warming, and so can accept a new image as easily as magnetic tape accepts a new audio signal. Again, the Post Office could almost abandon the old letter-post in favour of an instant 'Picfax-service' using the universal domestic TV as a receiver. Ideally, this would employ cheap overnight delivery on the telephone system, which is pretty dormant at night. As a minor benefit, TV addicts could peel off the screen Picfix 'stills' of their favourite stars or images. In fact Picfix Paper will become a universal medium for data and images, as temporary or as permanent as desired. And since you can write on it with a hot electrified point, it's an artistic medium too!

(*New Scientist*, 19 March 1981)

From Daedalus's notebook

Liquid-crystal compounds have two 'melting-points'. Below the lower one they are true solids; above the upper one they are true liquids; but between the melting-points they form a 'mesophase' liquid crystal which is a liquid with molecular order. For Picfix Paper, we want one whose lower m.p. is significantly above ambient (so that the image it holds is safely frozen) but easily reached by slight warming. 4-Methoxy-4'-(*n*-butyl) azoxybenzene (m.p.$_{low}$ = 42 °C, m.p.$_{high}$ = 77 °C) should do, at least for the pilot work.

To minimize the voltage needed to form the image, a 'twisted nematic' display is probably best; such displays can be flipped by 5–10 V, and less if we're clever. In the absence of the voltage, the liquid crystal molecules align themselves with a 90° twist between the faces of the display. Apply the voltage and they line up perpendicularly, erasing the twist. The normal way of making the molecular movement visible is to use crossed polaroids. For 'positive' Picfix we need the polaroids front and back to be in line. Then polarized light enters one side, is twisted by the liquid-crystal layer, and finds itself blocked by the other polaroid; so the paper will be 'normally black'. But the racing bright spot of the TV screen, accompanied by its voltage, will remove the twist and leave a corresponding local transparent 'bright spot' on the Picfix. For copying print, 'negative' Picfix might be better. It would have *twisted* polaroids so as to be 'normally white', and would be darkened by the spot, giving black-on-white printing.

Possible problem. Twisted nematic displays take tens of milliseconds to respond, much longer than the spot residence-time. We'll need to arrange the leakage conductivities so that the electrons dumped on the screen by the spot are discharged slowly over the 40 ms it will take for the spot to come back with new data. This will enable us to fix a single TV frame if we can cool the Picfix that fast. We should be able to make the thermal capacity low enough just by making the Picfix very thin — which is desirable anyway.

Uses. TV copying is the obvious one: video stills, Prestel, quick-look copy from computer terminals, etc. But there are at least two others:

(a) Thermograms. Lay the stuff on a surface, apply a quick voltage-pulse all over it, and you'll get a transparent image wherever the surface exceeds 42 °C (or whatever is the m.p.$_{low}$ of the liquid crystal). Peel off the paper and you've a permanent record. Useful for medical, engineering, insulation, etc., tests.

(b) Electrograms. Lay the stuff on a circuit, apply a brief pulse of heat all over it, and you'll get a transparent image wherever the voltage exceeds 5 V (or whatever is the $V_{threshold}$ of the paper). Again, peel it off and you've a permanent diagnostic record of the state of the circuit. Should be invaluable for tracing faults, finding wiring buried under plaster, checking batteries, seeing if cable is safe to touch, etc.

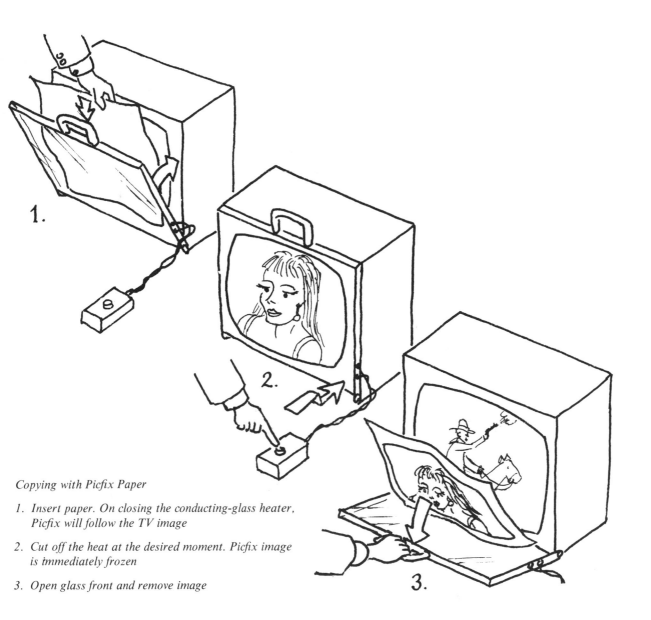

Copying with Picfix Paper

1. *Insert paper. On closing the conducting-glass heater, Picfix will follow the TV image*

2. *Cut off the heat at the desired moment. Picfix image is immediately frozen*

3. *Open glass front and remove image*

Non-Newtonian trousers, etc.

While struggling into a starched shirt recently, Daedalus began to ponder on the mechanical resistance of clothing. It occurred to him that fibres with strongly non-linear mechanical properties might be very useful in the garment business. The widest range of non-linear mechanical properties is shown by liquids, in the various forms of non-Newtonian viscosity. One extreme is dilatancy (very low resistance to shear at low shear-rates, but rapid thickening as the shear rate goes up) and the other is thixotropy or pseudoplasticity (resistance very high at low shear-rates, but collapsing rapidly as the shear-rate goes up). Various intermediate, offset, and hysteresis behaviours are also known.

So Daedalus is inventing ways of spinning hollow fibres filled with such liquids, and seeking fibre-forming polymer compositions which also show non-linear properties. His first goal is to produce strongly dilatant fibres. Textiles woven from them would yield readily enough to ordinary slow body movements, but would stiffen drastically against sudden exertions exceeding a set rate of flexing. They would be ideal for hyperactive children, self-destroying eager-beavers, and sufferers from nervous tics and incipient heart-attacks, because they would enforce a slow and graceful style of movement without sudden jerks or angularities. They should find a ready market in charm schools and yoga classes, where their encouragement of a smooth and elegant personal style would be much appreciated; they would also be issued to prison inmates to stop them running away. Unfortunately they would be very exhausting to fold and handle in the laundry, and their graceful flowing mode of deformation would resist the vigorous action of washing machines. So Daedalus hopes to exploit the temperature-dependence of dilatancy to produce garments which collapse to total floppiness at laundering temperatures, and recover when cooled down again.

Dilatant clothes would also have a useful protective function. As mandatory wear for football fans they would not only make it hard to throw a bottle or deliver a blow with much force; they would also protect the intended victim. His clothing would momentarily stiffen up or even lock rigid at the site of the blow, forming a sort of instantaneous local armour. Motorists too might thus be protected from the dangers of crashing, and at first Daedalus

imagined that the principle would be ideal for military uniforms. But he now realizes that the Army would prefer the exact opposite, and he is developing thixotropic fibres as a secondary goal. Uniforms woven from thixotropic fibres would resist all movement until stressed with sufficient vigour, when their resistance would collapse totally. Only when the movement ceased would they recover their ferocious stiffness. So the wearer would either be held completely stationary, or would be forced to move 'at the double' — the twin ideals of military life.

(*New Scientist*, 26 May 1977)

From Daedalus's notebook

How to fabricate non-Newtonian fibres? They'll have to be composite, with some conventional fibre-component to take the tensile loads — we don't want them sagging endlessly. So either a conventional fibre core must be coated with some rheologically appropriate composition, or a hollow-tube fibre must be filled with a suitable liquid. The second option seems better: we can then use non-Newtonian liquids as tacky or as runny as we like. We can fill the tubes in the same way that they put the soft centres in chocolates — use an initially solid core-composition which softens enzymatically after it has been coated with its outer layer. It would be nice to use some edible non-Newtonian confectionery mix as the fibre-core, but probably we'll have to devise a rather less biological piece of subsequent liquefaction-chemistry.

Manipulation of fibres. Weaving will be tricky, I suspect. We may have to use high enough temperatures to slump all viscosity to low levels. But the thixotropic fibres could be kept limp by high-speed vibration of the weaving and tailoring machinery: the clothes would only set up stiff when they were left static after completion. Incidentally, such clothes would have one neat military advantage, especially if their thixotropy showed marked hysteresis. Once 'broken' to some movement, their stiffness would be even more reduced for the precise retracing and repeating of the move. So after his first pace, a marching man will find that his otherwise unyielding trousers freely permit him to take more, identical paces. Precise repetitive military drill will become almost automatic.

The British flag is woven from DREADCO's thixotropic fibre. When buffeted by a strong breeze, its rigidity collapses and it streams gallantly in the wind. But as the wind dies away, it stiffens up and remains prestigiously extended while the flags of other nations droop incapably

The safety of obscurity

There is a clever photochromic glass, used in sun-glasses and even windows, which darkens on exposure to light. It works by a photographic-type breakdown of silver chloride into tiny grains of opaque silver. In dimmer illumination the reaction reverses and the glass becomes lighter; so it acts as a sort of automatic sunshade or venetian blind. In this connection Daedalus recalls the first principle of camouflage: to avoid contrasts. Many animals, for example, have dark backs and light bellies so that lit from above and shadowed beneath they appear unobtrusively uniform. Photochromic animals — frogs and chameleons, for example — go one better. They vary their overall colour to minimize their visibility, but even they cannot adapt each separate skin-area to its local illumination. Such a ploy would be so useful that Daedalus reckons Nature has already evolved it, but creatures thus perfectly camouflaged have so far escaped human observation.

So Daedalus intends to exploit the principle socially. He has long wondered why traditionally gaudy male garb became so drab in the Victorian era, and has remained so in business circles to this day. His theory is that social and sexual display became overcast by the need for camouflage in the capitalist jungle. One proven way of surviving in it is not to attract envious or disapproving attention: the eccentric or even the noticeable individual is asking for trouble. Hence the subfusc business suit and the city uniform. So Daedalus is developing a silver-chloride based photochromic suiting of unparalleled obscurity. Where the light shines on it, it will darken to reduce its reflectance; where illumination is low or shadowed it will lighten to reflect proportionally more light. Accordingly, its perceived brightness will be low and completely uniform. The human eye is so sensitive to contrast, and relies so heavily on it for interpretation of the visual world, that a garment thus rendered featureless by photochromic camouflage will be effectively invisible. Its wearer, deprived of the optical clues which might call attention to him, will not be noticed; photochromic hand-lotion and after-shave will complete the ultimate faceless organization-man.

Daedalus foresees great demand for his photochromic products not only among the lower echelons of the capitalist pecking-order, but also among sociologists, extra-marital adventurers, pick-pockets and plain-clothes detectives. And once most such people have attired themselves in DREADCO's artificial obscurity, the remainder will appear by comparison to be positively exhibitionist and will be forced — literally — to follow suit.

(*New Scientist*, 16 August 1973)

Daedalus comments

Silver-chloride glass exploits the archetypal photographic reaction: $AgCl \xrightarrow{h\nu} Ag + Cl$. In a photographic emulsion, the freed chlorine atom reacts irreversibly with the gelatine, leaving the silver atom as a nucleus for later development. In the glass, the chlorine can't get away, so the reaction is reversible. The balance of the photodissociation is set by the intensity of illumination of the glass.

Commercial silver-halide photochromics take minutes to react, and would be too slow to follow the rapid changes of illumination of the clothing of an active wearer. But more advanced photochromic systems — e.g. the spectacles which protect soldiers from nuclear-weapon flash — can darken in microseconds, and if adapted for clothing should fill the bill nicely.

The use the eye makes of shading in its evaluation of visual images is well discussed by J. Beck in *Scientific American* (August 1975, p. 62); an object which consistently disobeyed the normal rules of shading would certainly be hard, and maybe impossible, to evaluate. Even twisting and turning it in the light would not reveal its surface features; its surfaces would constantly change their shading to defeat scrutiny. But photochromism can go even further. Imagine a room decorated completely, walls, floor and ceiling, with photochromic paper. No matter how bright the lighting, any photochromic object placed in the room would be completely invisible — no contrast could arise between it and its background. A whole new field of illusioneering is clearly available to conjurors and camouflage artists!

'. . . creatures thus perfectly camouflaged have so far escaped human observation'

The stars wink down

Daedalus has been seeking a role for impecunious Britain in the expensive age of space-probe astronomy. We cannot hope to compete with the Americans, but with European aid we might just manage to push some fairly light and stupid object up to escape velocity. Solar eclipses, in which the Moon passes in front of the Sun, give a wealth of information. Daedalus therefore plans to orbit an opaque satellite to pass in front of the stars, thus producing equally informative 'stellar eclipses'. So DREADCO engineers are devising a tightly-folded polymer-film space-probe which will unfold in space under slight internal pressure to an opaque balloon 1 kilometre across. Launched into solar orbit in the plane of the Milky Way, it would have the right angular size to occult lots of interesting stars as it passed in front of them.

The great beauty of the scheme, says Daedalus, is that occultations can be observed without expensive high-resolution telescopes. The telescope needn't image the star at all well; it just has to intercept light from its general region — easily registered on a simple photomultiplier. Since stars shine steadily, a sudden change in intensity must mean an occultation. Its exact timing and extent give data on the position, size and radial brightness function of the occulted star more accurately than any telescope. Filters in front of the photomultiplier could give spectroscopic information too. For a balloon 1 km across, observers more than 1 km apart will see quite different occultations, so that a lot of amateurs with cheap telescopes (in the best British traditions) could rapidly gather a great deal of new knowledge.

At first Daedalus was afraid that we would have to ask the Americans to track our balloon accurately for us, and tell us where it was going to be next Thursday week. But he now reckons that aluminizing the balloon into a big convex mirror will make a tiny image of the Sun always visible in its centre, thus enabling it to be tracked visually. To avoid confusion, the aluminium will be coated with mahogany lacquer, giving the reflection a strong brown cast. On spectroscopic grounds there cannot be many brown stars, so identification should be unambiguous.

(*New Scientist*, 27 September 1979)

From Daedalus's notebook

We are interested in stars that can be identified as single objects by cheap telescopes, i.e. ones 3–30 000 light-years away (say $10^{16} – 10^{20}$ m) The typical diameter of a star is 10^9 m, so the range of angular diameters will be $10^{-7} – 10^{-11}$ radians. We should aim to put our balloon into a fairly elliptical orbit about the Sun so that the Earth's distance from it will vary from say 0.1–10 Earth–Sun distances i.e. about $10^{10} – 10^{12}$ m. So to be able to eclipse the stars of interest at some point in its orbit it should have a diameter around 10^3 m, giving it an angular size seen from Earth of $10^{-7} – 10^{-9}$ radians. Some of these eclipses will be neat and exact; for most the balloon will be too big and the event will be an occultation; for some it will be too small and the event will be a transit.

How to track the balloon? An aluminized convex balloon (like the early 'Echo' passive-reflectance satellite) would show a tiny image of the Sun near its centre. The Sun's apparent diameter is about 0.01 radians; its virtual image in the balloon will be smaller by a factor of about $r/2d$ where r is the balloon-radius and d is the distance of the Sun. So seen from about 1 Earth–Sun distance it will have an angular diameter about $\alpha = 0.01 \times r/2d = 0.01 \times 10^3/(2 \times 10^{11}) = 5 \times 10^{-11}$ radians; i.e. much the same as those of the various stars being occulted by the balloon. So the solar image will be (a) easily visible and tracked but (b) not so bright as to swamp the radiation from target stars.

Data-collection. A cheap telescope is aimed at the star of interest with a photomultiplier gathering light from just that region. A big mirror or lens with high aperture but no attempt to minimize aberrations, is all you need. Something like a searchlight mirror would do (Hanbury-Brown and Twiss used such mirrors with photomultipliers in their famous work on intensity-fluctuations in Sirius). The optics needn't image the star well, or even resolve it from neighbouring unocculted stars, which will only raise the steady base-line of illumination seen by the photomultiplier. You just record the variation in photomultiplier output with time, and look for the characteristic dip-pattern as the target star is occulted. One attractive scheme — elaborate the balloon a bit. Paint black patches on it and give it a spin. Then the Sun's reflection will wink on and off to give a time-base. Even better — give it windmill arms! Then (1) it will sweep a lot more sky and provide many more occultations and (2) with proper arm-design stars of different diameters would give very different occultation signals. Angular-diameter measurements for many stars could then be rapidly accumulated by the nation's amateurs.

Incidentally, there's a lot to be said for putting another occultation satellite into close orbit around the Earth. Its larger angular diameter would create many more occultations, even though they'd be briefer. And in polar orbit it could cover the whole sky.

Reflection of sunlight in spherical balloon

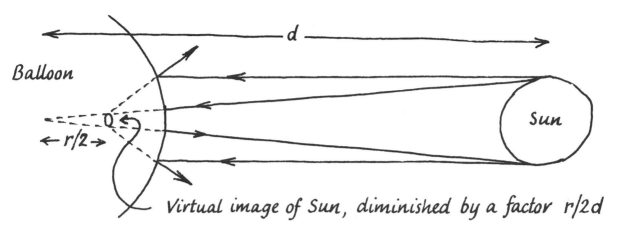

Balloon

$\leftarrow r/2 \rightarrow$

d

Sun

Virtual image of Sun, diminished by a factor $r/2d$

Occultation satellite

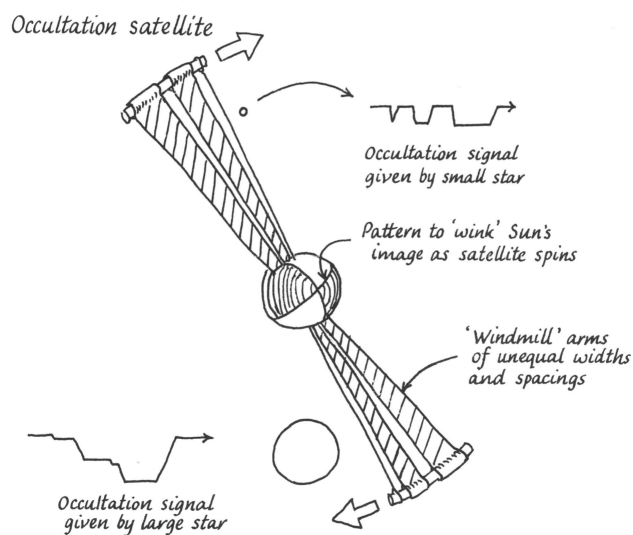

Occultation signal given by small star

Pattern to 'wink' Sun's image as satellite spins

'Windmill' arms of unequal widths and spacings

Occultation signal given by large star

Boats, boots, and boiling water

Skin-friction in ships rises almost as the cube of the speed, and grows steadily worse as things like barnacles accumulate on the hull. So Daedalus has been musing on the enormous savings that would result if it could be eliminated. He first thought of copying the hovercraft air-cushion principle, by making the ship's hull from porous metal and keeping it pumped up. The thin film of escaping air would provide perfect lubrication for the moving hull, but if the pumps failed the ship would slowly fill up and sink. Daedalus then recalled the 'spheroidal effect' which prevents a water-drop touching a red-hot plate, so that it skitters about on a cushion of steam. A red-hot vessel would therefore generate its own self-maintaining steam cushion — and the barnacle problem would also be neatly overcome at the same time. The power requirements could be quite modest, for heat-transfer actually drops quite sharply at spheroidal temperatures (the water can no longer touch the metal to cool it). But corrosion problems would still be severe, and effective insulation would be needed if life were not to be the traditional hell below decks. Daedalus is studying a conversion of the common two-skin hull into a vacuum-jacketed design, effectively a floating Thermos flask electrically heated on the outside. While red-hot propellers would probably be unusually efficient, Daedalus's aesthetic sense prompts him to suggest an integrated power-plant for his hot ship. He proposes internal heated water-ducts to take in seawater from the front and expel steam in blasts from the back. His underwater steam pulse-jet is of course merely a greatly scaled-up version of the toy pop-pop boat.

(*New Scientist*, 25 May 1967)

The Central American basilisk lizard (known locally as the Jesus Christ lizard) can actually walk on water. It does this by rapid pedalling with its wide, lobed feet, rather as a flat stone can skip on water by force of impact. If a lizard can do it, says Daedalus, so can a man! But after a few abortive experiments in which volunteers wearing adapted snow-shoes tried to run a length of the DREADCO Recreation Club swimming baths, he has been forced to take the technical challenge of water-walking more seriously. It needs a shoe of wide area that will impact on the water but not sink in far enough to require forceful dragging out. The simple solution is to make the shoe red-hot; then it would never touch the water at all, but would be instantly cushioned on a layer of steam. The steam-pressure would be mainly upwards, but by slight angling of the foot a rearward propulsive jet could be expelled, giving to the water-walker a high degree of power-assistance. So Daedalus's prototype water-boot has a wide flange-sole separated from the foot and heated to redness by a small propane burner. The 'water-bootman' strides smartly across the liquid surface, generating a dramatic hiss at each footfall. The combination of steam power and frictionless gas support gives him a top speed of many knots. Indeed, the slipperiness of the water-surface capsized early volunteers, who sank with a fizz and a splutter, so Daedalus has modified his design by adding centreboards to guide the boots skate-like on the water.

Thus the deep loses much of its terror. Victims of coastal shipwreck will simply be able to walk ashore, risking neither drowning nor freezing to death; in fact overheating will be the main problem. Intrepid water-bootmen will explore the new sports of oceaneering and ocean mountaineering, negotiating choppy seas or huge swells, while novel and most unfair ways of catching waterfowl will become possible. And maybe basilisk lizards will become popular pets, for the sheer delight of taking them for a stroll on the local pond.

(*New Scientist*, 10 January 1974)

Daedalus comments
Joshua Laerm's high-speed film studies of the basilisk lizard are described in *Scientific American* (September 1973, p. 70).

From Daedalus's notebook

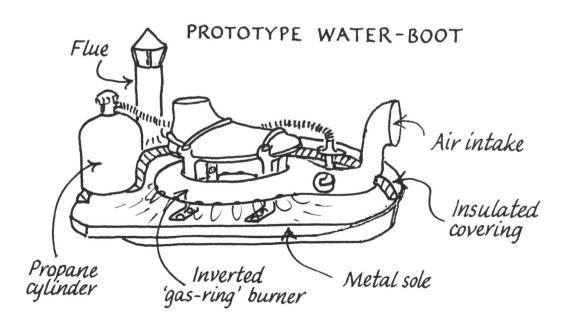

PROTOTYPE WATER-BOOT

Flue

Air intake

Insulated covering

Propane cylinder

Inverted 'gas-ring' burner

Metal sole

Bicycles made for the zoo

Daedalus criticizes current attempts to communicate with dolphins, pointing out its impossibility without common subject-matter. He therefore intends to introduce the creatures to the dry world by a form of inverted aqualung. This is a framework to support the dolphin, suspended from a small autoballasted balloon and fitted with recirculating skin-dampening pump and water-filled goggles (to correct refraction problems). Just as the human swimmer wears flippers, so the dolphin will locate its fins in large air-vanes giving it traction through the less-viscous air. The animals will gain immensely from the challenge of a new environment — Daedalus believes their great brains are stagnant from the boredom of their homogeneous natural habitat. He envisages them exploring the world freely and then diving back into their pool, when the autoballasting system will take up enough water to leave the vehicle 'parked' for later use. To get a feel for the problems involved, Daedalus's prototype system is simply a wheeled tank running on a closed-loop rail-track, part of which runs submerged through the dolphins' pool. The dolphins will soon learn to enter the tank and take a trip round the track, thus acclimatizing to the idea of controlled movement through the dry world.

Another similar project is concerned with draught animals, which still provide most of the motive power in undeveloped countries. Daedalus greatly admires the bicycle, which increases human mobility at least fivefold by good muscular matching and releasing the leg-power wasted in supporting oneself. An equivalent 'buffalo-bike' would increase enormously the useful work available from these powerful but low-geared creatures. Daedalus's four-wheeled prototype (no balancing needed) is driven by an automatic gearbox from treads worked by the animal, which is firmly supported by undergirding. A human steersman is needed, but Daedalus is making a fully self-controlled model to see how buffalo respond to such a novel extension of their abilities. He fears an increase in aggressiveness paralleling that of human drivers, and a disinclination to move henceforth by normal means.

(*New Scientist*, 25 September 1969)

From Daedalus's notebook

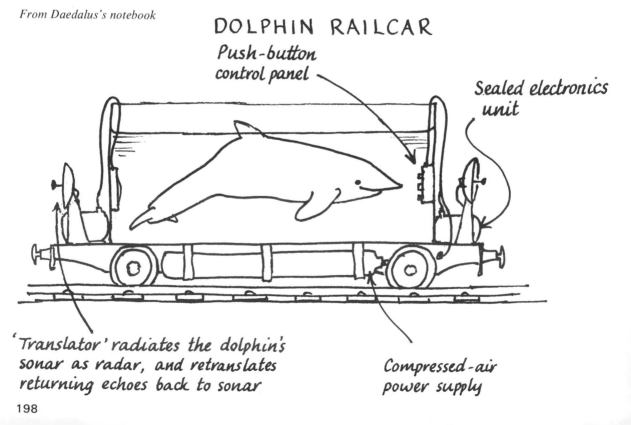

DOLPHIN RAILCAR

Push-button control panel

Sealed electronics unit

'Translator' radiates the dolphin's sonar as radar, and retranslates returning echoes back to sonar

Compressed-air power supply

BUFFALO - BICYCLE

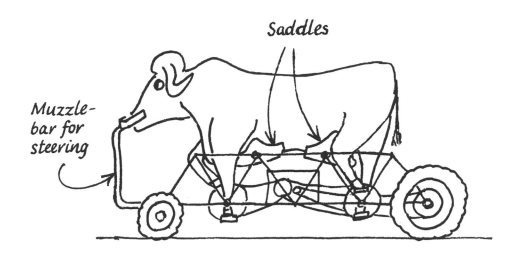

Saddles

Muzzle-
bar for
steering

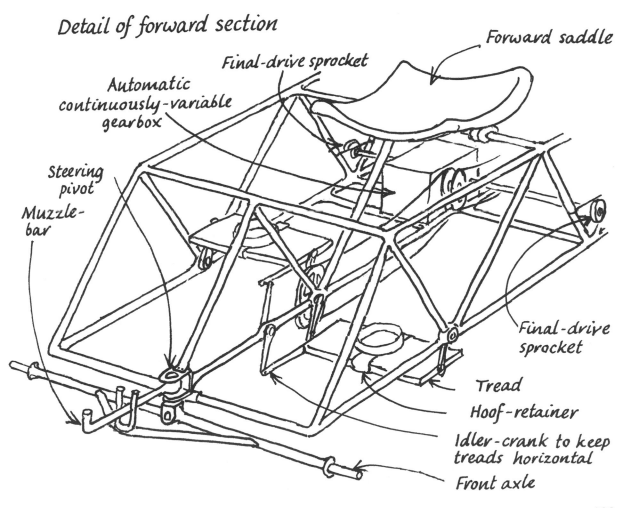

Detail of forward section

Forward saddle

Final-drive sprocket

Automatic
continuously-variable
gearbox

Steering
pivot

Muzzle-
bar

Final-drive
sprocket

Tread

Hoof-retainer

Idler-crank to keep
treads horizontal

Front axle

Subtle approaches to family planning

Daedalus shares the current widespread alarm about the frightening rate of increase of the world's population. Unfortunately, a major contribution to this deafening thunder of tiny feet comes from those too ignorant, feckless, or poorly motivated to use contraceptive techniques successfully; and Daedalus now proposes a quite novel solution to this thorny problem. DREADCO pharmacologists are at work devising an *addictive* contraceptive pill. The present prototype contains about the most harmless of effective addicting agents, nicotine. Even the dimmest schoolgirl or most harassed mum could no more forget to take it than she could accidentally forget to smoke! Pregnancy would require as firm an effort of will as its avoidance does at the moment, so 'unwanted' children would simply never occur. Daedalus is also devising compositions with a little more kick to them, which will be introduced into the pop scene by DREADCO pushers, thus reaching a segment of society peculiarly at risk. With any luck, a high proportion of the young will be hooked as they pass through the 'experimenting-with-drugs' phase; Daedalus hopes to get his compositions denounced in the media as particularly immoral so as to be sure of a large market among rebellious youth. In this way the future population will stabilize and even, mercifully, begin its long decline to a sensible level. Indeed, controlling the supply and price of such a pill would be an ideal way of implementing a government population policy without overt compulsion. The government makes millions out of nicotine addiction as it is, without any corresponding social benefit! Maybe the best way to translate the idea into practical politics would be to invert it and create a contraceptive cigarette.

(*New Scientist*, 10 December 1970)

Daedalus comments

It didn't take long for this idea to be appropriated, and to get into circulation at least as an ideal against which other contraceptive strategies could be compared. About a year later it surfaced in an article in *The Financial Times* (2 December 1971, p. 22): 'Growth and anti-Growth' by Lord Robbins. Discussing the value of increased productivity and education in decreasing the reproductive ambitions of the average family, he commented: 'The population problem is not going to be solved by the degree of increase of productivity per head that it is rational to hope for; only the invention of a Pill as habit-forming as alcohol or tobacco can do that now.'

But so far, despite the enormous attractions of a product combining the impeccable moral status of a personal and social benefit with the security of a firmly addicted market, no pharmaceutical company to my knowledge has yet taken up the challenge.

Fertilization in mammals is made easier by a 'sex-attractant' exuded by the ovum to attract the sperm, which stand a very low chance of meeting it by random exploration. As it is, they need only swim up the concentration-gradient established by the attractant to be fairly sure of finding the ovum at the end of it. Nature is very economical in her methods, so Daedalus reckons this substance is probably a steroid or fatty acid like the sex-attractant odours used by female mammals to attract their mates. If so, it should be quite easy to identify and synthesize, and makes possible the cunning new contraceptive ploy of misdirecting the sperm. An appropriately placed spray of sperm attractant will have them all clustering vainly round the lady's knees, say, or winding up under the pillow. But Daedalus goes further, and points out that the vagina normally harbours a stable assortment of harmless microorganisms: specialized yeasts and so forth. Already bacterial cultures have been used to synthesize chemical precursors for contraceptive-pill steroids, and Daedalus had the idea of trying to develop this synthetic ability in human vaginal yeasts. A stable vaginal yeast population which continually elaborated contraceptive steroids would of course be an ideal long-term form of birth control requiring no attention or knowledge on the part of its user. But Daedalus now reckons it should be easier, and far more fun, to teach the yeasts to synthesize sperm-attractant, which has the added advantage of being an entirely natural body product and therefore free of side-effects.

So DREADCO biologists are culturing vaginal yeasts under mutagenic radiation, feeding them basic steroids to play with, etc., to create a strain which synthesizes sperm-attractant. Even unrelated microorganisms can exchange plasmids carrying genetic information, so once one organism has learnt the trick, it should be able to 'teach' all the others. And once these harmless organisms have taken up residence in the vagina, and multiplied to their stable population of some billions, effective contraception will be well assured. The yeasts will divert and seduce the sperm away from their high purpose, leaving the real ovum with practically no chance of attracting attention amid the horde of alluring impostors.

This subtle contraceptive technology will halt the population-explosion at a blow. Once established, the new yeasts will flourish in their hostess indefinitely, so that to have a child she must obtain a prescription for a special antibiotic to quell them for a time. Furthermore, the yeasts should spread through the population in epidemic fashion, so that fertility will drop universally. And the feckless heavy-breeders who perpetuate the cycle of deprivation at the bottom of the social scale will now be too feckless to seek the medical advice which alone could restore their sabotaged fertility.

(*New Scientist,* 3 April 1980)

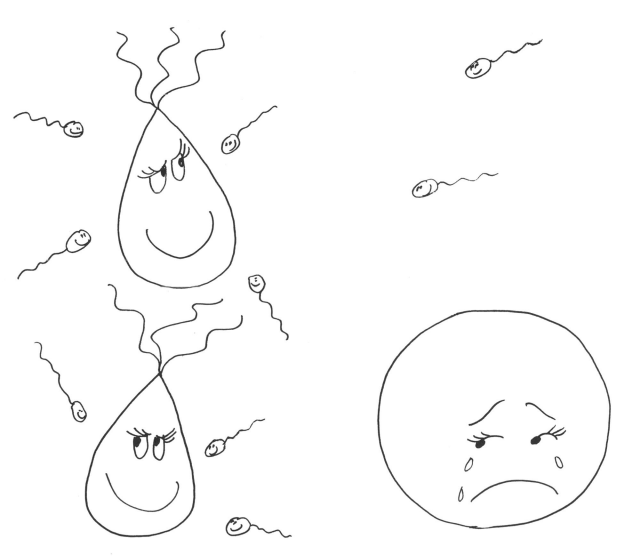

Constructive uses of adulteration

Like many other non-smokers, Daedalus objects to being forced to breathe other people's tobacco-smoke. In self-defence he has devised a simple anti-smoking ploy. Many relatively harmless substances are oxidized in a flame to poisonous products: thus carbon tetrachloride vapour gives the highly irritating phosgene. Carbon tetrachloride itself is rather toxic, but is closely related to the odourless and inert halocarbons used as refrigerants and aerosol propellants. So Daedalus's anti-smoker gadget is a little aerosol which can be sprayed surreptitiously into the air of (e.g.) a railway carriage. Non-smoking passengers will notice nothing, but the deviant tobacco-combustor will find himself inhaling such beastly decomposition-products that he will quickly give up. As well as being ideal for such personal defence, this neat device might also be used to enforce non-smoking regulations in, e.g., cinemas, or airliners during take-off.

This principle could easily be extended to other chemically fraught social situations. Thus Daedalus recalls antabuse, the drug which interferes with alcohol metabolism so as to make the heavy drinker excessively hot, bothered, and disinclined to imbibe further. It should be easy to devise an antabuse-doped peanut for cocktail parties, the judicious deployment of which would enable hostesses to keep matters under control without overt authoritarianism. And in the same vein Daedalus points out how critical are small concentrations of key flavour-substances in giving food its palatability. He is seeking tasteless compounds to react with the flavouring in, e.g., doughnuts and cream cakes, so as to subvert their taste and make them seem quite horrid. The weak-willed weight-watcher could then take Daedalus's anti-cake pill just before a meal, and be immunized effortlessly against her baser compulsions. Unfortunately the ruthless competition of the catering world could easily result in rival café proprietors doping each other's water or atmosphere with these gourmet-sabotaging products, to the general discredit of *haute cuisine.*

(*New Scientist,* 10 May 1973)

In a search for possible National Health Service economies, Daedalus once advocated replacing the entire GP network by an annual issue of a dozen undated sick-notes to each member of the work-force and a large number of slot machines dispensing pills compounded of a mixture of aspirin, penicillin and valium. He now acknowledges that self-medication, while it could well do for medicine what the supermarket did for grocery, does make overdoses dangerously easy to obtain. He intends to counter this in the same elegant way that society has tamed those powerful drugs caffeine and alcohol: high dilution. It is possible to damage yourself by drinking too much coffee or beer — but limited bladder-capacity saves most of us from overdosage. So DREADCO bio-chemists are devising novel foods and drinks incorporating small percentages of pharmaceuticals: a 'nightcap' drink which contains a genuine soporific, an antibiotic sausage (which incidentally resists bacterial spoilage), a breakfast cereal spiked with benzedrine which actually does get you off to a brisk start in the morning. To ensure that these creative products are not misused, they will resemble coffee and beer in having a taste which is rather off-putting to the naive palate. Only while they relieve your symptoms will you feel a 'yen' for them, just as you unconsciously learn to value coffee and beer for the subtle mental alleviation they bring about. The new products will compete in the shops at high prices; not, explains Daedalus, to maximize DREADCO profits, but to reinforce their odd taste by deterring those who won't benefit from them. Daedalus has great faith in people's ability to learn instinctively what's good for them, provided they are not seduced by low prices or appealing flavours. But to prevent long-term dependence or misuse of his 'Pharmafoods' (Regd) he will switch the formulations around from time to time: e.g. by transferring the soporific from the nightcap drink to the sausage. Purely habitual nightcaps users will thus be freed from long-term medication, although they may continue to benefit from the placebo effect. But the real insomniacs will soon unconsciously pick up the bedtime sausage habit.

(*New Scientist*, 19 February 1976)

DREADCO's anti-smoker spray

Index